Advances in Geographical and Environmental Sciences

Advances in Geographical and Environmental Sciences synthesizes series diagnostigation and prognostication of earth environment, incorporating challenging interactive areas within ecological envelope of geosphere, biosphere, hydrosphere, atmosphere and cryosphere. It deals with land use land cover change (LUCC), urbanization, energy flux, land-ocean fluxes, climate, food security, ecohydrology, biodiversity, natural hazards and disasters, human health and their mutual interaction and feedback mechanism in order to contribute towards sustainable future. The geosciences methods range from traditional field techniques and conventional data collection, use of remote sensing and geographical information system, computer aided technique to advance geostatistical and dynamic modeling.

The series integrate past, present and future of geospheric attributes incorporating biophysical and human dimensions in spatio-temporal perspectives. The geosciences, encompassing land-ocean-atmosphere interaction is considered as a vital component in the context of environmental issues, especially in observation and prediction of air and water pollution, global warming and urban heat islands. It is important to communicate the advances in geosciences to increase resilience of society through capacity building for mitigating the impact of natural hazards and disasters. Sustainability of human society depends strongly on the earth environment, and thus the development of geosciences is critical for a better understanding of our living environment, and its sustainable development.

Geoscience also has the responsibility to not confine itself to addressing current problems but it is also developing a framework to address future issues. In order to build a 'Future Earth Model' for understanding and predicting the functioning of the whole climatic system, collaboration of experts in the traditional earth disciplines as well as in ecology, information technology, instrumentation and complex system is essential, through initiatives from human geoscientists. Thus human geosceince is emerging as key policy science for contributing towards sustainability/survivality science together with future earth initiative.

Advances in Geographical and Environmental Sciences series publishes books that contain novel approaches in tackling issues of human geoscience in its broadest sense—books in the series should focus on true progress in a particular area or region. The series includes monographs and edited volumes without any limitations in the page numbers.

Deepak Kumar

Renewable Energy Scenarios in Future Indian Smart Cities

A Geospatial Technology Perspective

Springer

Deepak Kumar
Center of Excellence in Weather
and Climate Analytics
Atmospheric Sciences Research Center
(ASRC)
State University of New York (SUNY)
Albany, New York, USA

Amity Institute of Geoinformatics
and Remote Sensing (AIGIRS)
Amity University Uttar Pradesh (AUUP)
Noida, Uttar Pradesh, India

ISSN 2198-3542 ISSN 2198-3550 (electronic)
Advances in Geographical and Environmental Sciences
ISBN 978-981-19-8455-6 ISBN 978-981-19-8456-3 (eBook)
https://doi.org/10.1007/978-981-19-8456-3

This Springer imprint is published by the registered company Springer Nature Singapore Pte Ltd.
The registered company address is: 152 Beach Road, #21-01/04 Gateway East, Singapore 189721, Singapore

Preface

Urban renewable energy generation is hampered by limited installation area. As a starting point for an analytical framework for decarbonized urban settings, we use the equilibrium between city energy demand and renewable energy density. Space-based solar energy, small modular nuclear power systems, deep geothermal systems, and other producing options may revolutionize the energy landscape in waves of innovation. These strategies may reduce the demand for fresh water, clean air, and other resources and greenhouse gas emissions. Large-scale use of renewable energy is crucial for satisfying urban energy demand and reducing emissions in today's cities. As technology advances, renewable energy will become more efficient, user-friendly, cost-effective, accessible, and sustainable.

Both industrialized and developing countries must innovate in energy and practice responsible consumption due to the global energy crisis and climate change. 2030 energy demand is expected to climb 60–85%. IPCC recommends no more than 450 ppm of greenhouse gases in the atmosphere to limit global warming to 2 degrees Celsius above pre-industrial levels (ppm). NASA claimed 400 ppm was attained in March 2015. If the world wants a successful, healthy, and environmentally safe future, it needs another industrial revolution fuelled by easily accessible, low-cost, and eco-friendly energy. Some rich nations have decoupled economic growth from energy usage to reduce resource use and environmental impacts. This was achieved by utilizing waste heat to create power. Decarbonizing energy sources and enhancing energy efficiency are key.

Renewable sources are the sole realistic choice moving forward, notwithstanding fossil fuels' ongoing relevance in urban energy generation. Cogeneration and district heating, which are fuel-efficient, may be important in metropolitan areas. As cities develop, renewable energy becomes "energetically important." Changing the energy source is just the beginning; the transition will be effective provided the new source is economical, long-lasting, and growth friendly.

Solar power doesn't need any water, eliminating environmental concerns about increased water demand and shortages. Recent advances in solar technology (particularly concentrated and photovoltaic solar power) have cut their implementation costs, making them competitive with fossil fuel-based power generation in mid- and high-latitude locales. Solar photovoltaic electricity grew 58% annually between 2006 and 2011, making it the fastest-growing renewable technology. Concentrated solar power rose 37%, while wind power rose 26%. Solar panels and photovoltaic materials may be mounted on building roofs with minimal aesthetic effect, making them ideal for urban areas. Future developments in on-site renewable energy production may allow for zero-emission constructions and energy-efficient low-carbon eco-cities. Every day, new technology makes cities greener and more self-sufficient. Wind, solar, and rainfall harvesters are being developed for high-rise buildings to enhance energy production. It reduces dangers connected with urban wind turbines. Technological advancements may increase eco-cities. These urban neighbourhoods have water-efficient fixtures, double-paned windows, a southerly orientation for passive heating, solar photovoltaic roofs and walls, and power-generating stations.

The incompatibility of supply and demand, as well as their incorporation into the energy system, might pose challenges to the widespread use of renewable energies in urban settings. With the help of smart grids, electricity distribution might be more efficiently regulated. Improved energy security and dependability; decreased transmission costs; use of existing infrastructure; and reduced land requirements are only some of the advantages of implementing such measures in an urban setting. Since the early 1980s, various nations have successfully implemented renewable energy systems for residential applications, proving the systems' maturity as a technology. The industry would benefit from international or regional certification methods that provide high standards without ignoring local customs and customs regulations. With the support of long-term mandates, the energy market has developed into one that is both stable and lucrative. In an effort to lower home energy expenses, provide employment opportunities, and keep up with the ever-increasing renewable energy demand, the government has begun promoting tiny, residential solar water heating systems among low-income households or social housing complexes. Additional appealing alternatives include the provision of low-interest financing and the establishment of solar-based district heating. Sustainable cities must be low-carbon, resilient, and liveable, either in themselves or in their resources. Climate change, harsh weather, and natural and human-caused disasters threaten metropolitan energy infrastructure. The energy infrastructure of a city also affects residents' health and happiness. Transport, food production, and water quality are high-energy businesses that influence health, economic competitiveness, cultural appeal, and social, gender, and racial equality. Decentralized, non-fossil fuel energy generated in urban areas may help cities meet sustainability objectives. Using renewable energy choices redundantly makes areas more self-sufficient and resilient. Greater power densities of renewable energy technologies, improved infrastructure capable of supporting vast integrated energy producing systems, and better urban energy efficiency, particularly in buildings, may help overcome these challenges. Recent technological breakthroughs might make cities more sustainable. The environmental, economic, and social elements of urban sustainability might all benefit greatly from the implementation of the city-integrated renewable energy idea, which involves generating energy where it is really needed. Benefits of such distributed energy systems include the following four features: (i) carbon-neutral products and services, (ii) decrease the need for costly network upgrades, (iii) facilitate decentralized power generation and communication, and (iv) stimulate social resources and unity.

Albany, New York, United States of America (USA)

Deepak Kumar

About This Book

This book presents recent advances in renewable energy scenarios for future Indian smart cities including technologies and devices at the scales of both experimental and theoretical models for Industry 4.0, the concept of automated and computerized industrial manufacturing and practices. The current Indian economy is inclined towards smart urban cities, but the energy deficit in modern society is not well recognized. As a result, there is an enormous need to explore alternative avenues of energy for future smart cities. Because such cities depend significantly on technologies and devices that comprise Industry 4.0, the synthesis of energy scenarios enables an understanding of the technology, applications, and devices that contribute immensely to the textile, construction, cosmetics, biomedical, and environmental industries, among others. These industrial areas are the key starting points for a wide range of applications, consequently becoming top priorities for science and technology policy development. Such advances already have been adopted in various contemporary services and products, especially in the fields of electronics, health care, chemicals, cosmetics, composites, and energy. This book is a valuable resource for practising energy planners, citizens, and professionals such as businesspeople, bureaucrats from all levels of government, employees from non-governmental public organizations and their volunteers, and other individuals who have stakes in the development of their city region.

Contents

About the Author

Dr. Deepak Kumar is an academic research professional with extensive experience of over 10 years in the field of remote sensing, analytical research methods, geospatial technology applications, and urban planning. He is presently working as a Research Scientist at the Center Of Excellence in Weather and Climate Analytics, Atmospheric Science Research Center (ASRC), State University of New York (SUNY), University at Albany (UAlbany) Campus, Albany, New York (USA). He has been associated with Amity University Uttar Pradesh, Noida (India), as an Assistant Professor since 2016. He has been revisiting a wide range of issues associated with traditional research activities but has his own ingest of thinking on other non-traditional/challenging areas/topics for the exchange of research activities. More preciously, his teaching and research interests lie around the intersection of Geospatial Sciences, Sustainable Climate Sciences, and Computational Sciences for Smart Cities. He has completed two government-sponsored research projects as a sole principal investigator, namely "Hybrid Urban Landscape Analysis for Green Smart Cities through Geospatial Technology" sponsored by the Science and Engineering Research Board, Department of Science and Technology, Government of India, and "Meta-sensing of the urban footprint from airborne synthetic aperture radar (ASAR) data" sponsored from Space Applications Centre (ISRO), Ahmedabad, Gujarat, India. As principal investigator of government-funded R&D projects, Dr. Kumar has led a cross-functional research team. He has been involved in project development, management, laboratory setups, supervision, along with academic research training to undergraduate, postgraduate, Ph.D. Students and fellow researchers. His research findings are published in reputed peer reviewed international journals indexed by Elsevier, Springer, Taylor & Francis, Emerald, SAGE, Wiley etc. He has reviewed more than 150+ research articles for the reputed high-impact factor journals as an ad-hoc/ adjunct/ invited reviewer or editorial board member. He encourages researchers to develop their skills, knowledge, and experience, in aspects of academic research through research teaching, conference appearances, outreach activities, and contribution of professional membership of learned bodies.

The responsible series editor of this book is Dr. R. B. Singh.

Chapter 1
Concept of Sustainable Energy System for Smart Cities

Abstract Energy fuels the life on the earth. All the fundamental components existing in this biosphere needs energy for their survival. Everything, directly or indirectly, depends on the sun's energy. Humans and animals need the sun's heat and light for surviving. Plants need it to conduct the process of photosynthesis. Later when the remanent of these animals and plants is fossilized for millions of years, it results in the creation of various types of fossil fuels depending on the type of fossil, the amount of pressure and heat. These fossil fuels are a great source of energy, but they are finite, the time taken for their formation is a lot as compared to the rate at which they are being used in day-to-day life, making them a non-renewable source of energy. Using fossil fuels like burning coal, using crude oil, petroleum, and natural gas leads to the release of greenhouse gases, giving rise to climate change and environmental degradation. Apart from non-replenishable energy, there is renewable energy derived from renewable sources of energy, as they are naturally replenished on a human timescale like solar energy, hydro energy, biomass energy, geothermal energy, tidal energy, and wind energy. Renewable energy is also referred to as clean energy because of its low carbon emission quality. Energy conservation and efficiency are becoming major goals, the world needs to work upon. A lot of energy is wasted on daily basis, which is to be conserved, not only for cutting costs but also for sustainable use of our natural resources.

Keywords *Energy · Energy efficiency · Renewable · Energy intensity*

1.1 Introduction

The escalating development of our society is the result of the human ability to transform energy from one form to another. In this transformation, some amount of energy is wasted. Decreasing the amount of this wastage of energy by using efficient machinery is a step towards energy efficiency (Oral et al. 2021; Rn et al. 2014). Energy consumption has only increased forever. No one can be blamed for this increasing trend other than the increasing urbanization, unsustainable development, industrialization, transportation, manufacturing, and many others. With this increasing global energy consumption, using energy efficiently and sustainably is a

D. Kumar, *Renewable Energy Scenarios in Future Indian Smart Cities*, Advances in Geographical and Environmental Sciences, https://doi.org/10.1007/978-981-19-8456-3_1

must, otherwise, mankind will have to face harsh effects some of which are visible and alarming like climate change, loss of biodiversity, abnormal rapid melting of the glaciers, depleting ozone layer, adverse effects on public health (Chen et al. 2000; Khan et al. 2021). Earlier, renewable energy has been used to do daily jobs like transportation lighting, heating, and more. The wind was used for sailboats and the waves. The power of the wind was used to grind grains by windmills (Bataineh 2012; Iqbal and Iqbal 2019; Tanskanen et al. 2021). The traditional biomass was used for various purposes. Humanity, however, has spent the last few thousand years shifting away from cleaner and more expensive energy sources like solar and wind in favour of coal (Kaur et al. 2020; Riggs et al. 2017). The usage of renewable energy, from residential rooftop solar panels to industrial-scale wind farms, is growing as more efficient and cheaper methods of storing this kind of energy become available (Ramirez et al. 2015; Rustemli and Dincer 2011). In some cases, entire rural communities depend on these resources only for heating and lighting. All types of energy have various environmental effects, but non-renewable energy sources are substantially more harmful than renewable sources (Gambhir et al. 2010; Shah et al. 2019; Yamin and Bhat 2021). However, renewable sources like geothermal, hydropower, and biomass also have significant environmental impacts (Bellezoni et al. 2021; X. Li et al. 2021a, b; Yadav et al. 2019). The type and degree of environmental impacts caused depend on various factors like the technology used, the geographical location, and many others. The intensity and the potential, of these environmental impacts involved with the use of renewable energy, holds must be understood, so that they can be judiciously used to avoid or minimize these impacts.

In recent years, energy sustainability in cities has been actively explored and incorporated into the UN Sustainable Development Goals (Bian et al. 2020; Li et al. 2020; Yaya et al. 2016). No appropriate approach has been proposed to quantify the complicated chain of responses affecting urban energy infrastructures (Jin et al. 2019; Kumar 2019; Lemlin et al. 1997). Physical interactions between urban buildings affect energy consumption in cities. Different approaches have been taken to energy efficiency and city sustainability. Extrapolating basic urban structures to present cities' complex urban morphologies is not advisable (Tan et al. 2016; Thomas et al. 2008; Yang et al. 2021). Aggregated urban formations may reflect complex urban morphology behaviour. Urban aggregates react differently than modular ones. Similarities include energy demand, renewable energy integration, and system design. Adding a neighbourhood to a crowded city has minimal influence on its complexity. When introducing new structures on the city's periphery or in the suburbs, the neighbourhood plan should be carefully considered. Urban shapes have a higher financial impact on the energy system than annual energy or peak demand, according to this study. Urban energy simulators can't properly depict urban forms. Urban form influences cost, autonomy, and renewable energy integration (Aziz et al. 2018; Khanna 2012; Pramanick et al. 2021). Future city energy transitions need urban planning and energy system optimization. Climate conditions, building restoration, and energy technology costs may significantly affect quantitative outcomes.

Urban areas need a lot of power because of the density of the population there. These needs are often beyond the capacity of the local economy to provide (Adelaja et al. 2011; Leslie et al. 2020; Papaefthymiou and Dragoon 2016). Options for meeting these needs that have been established locally might vary even within the confines of a single community. European towns are encouraged to cut energy use, boost renewables, and decrease reliance on fossil fuel imports as part of the continent's governing philosophy (Hardiman et al. 2017; Msheik et al. 2021). Numerous cutting-edge ideas and tools exist to meet these requirements. In this study, we provide a systematic approach to the planning and evaluation of the evolution of "smart" energy systems, which will inevitably include intricate networks of technologies for the supply of energy that draw upon a wide variety of local and remote sources (Aziz et al. 2018; Colmenar-Santos et al. 2019; Tcholtchev and Schieferdecker 2021).

Smart city design considers both economic and environmental sustainability. A comparison of building construction and operation, urban infrastructure, and transportation shows that energy is a crucial role in the ecological impact of human settlements (Colmenar-Santos et al. 2019; Khanna 2012; Perez et al. 2016). According to the Sustainable Process Index, two-thirds of a smart city district's environmental impact may be attributed to energy sources (most of it for domestic heat purposes). By using industrial waste heat instead of fossil fuels, we can benefit the environment (Aziz et al. 2018; Marco et al. 2006; Perez et al. 2016). This might boost regional value creation by reducing fossil fuel imports.

1.2 Defining Energy

According to scientists, energy is the capacity of doing work. There are various forms in which energy exists, such as thermal energy, electrical energy, gravitational energy, light energy, chemical energy, nuclear energy, etc. (Balasubramanian and Ariffin 2014; Basir et al. 2013; Du et al. 2022; He et al. 2021). These energy types revolve around motion, say a body in motion possesses kinetic energy, whereas a body at rest possesses potential energy. According to Sir Albert Einstein, energy remains constant always, it can neither be destroyed nor created, but it can be converted from one form of energy to another, and this is called the 1st law of thermodynamics or the conservation of energy (Borozan and Pekanov Starcevic 2018; Dover 2018; Rowe 2016). Say an apple hanging on a tree possess potential energy and as soon as it gets detached, as it falls its potential energy is gradually converted into kinetic energy (Lee et al. 2021; Saraswat et al. 2021). Also, nuclear energy is a kind of potential energy as it is the result of the configuration of subatomic particles in the nucleus of the atom. Similarly, the kinetic energy of water is harnessed and converted into electrical energy, which is then converted into light and heat (Lai and Locatelli 2021). The chemical energy of food gives energy to humans for doing work by converting it into kinetic energy.

1.3 Forms of Energy

1. **Potential energy** is also referred to as accumulated energy and the energy due to position as it is the energy by the virtue of the position of an object which depends upon the relative position of the parts in the system.
2. **Chemical energy** is also stored energy. It is about how atoms and molecules are put together and how their bonds store energy. Chemical energy can be found in batteries, biomass, oil, natural gas, and coal. When wood is set on fire in a fireplace, its chemical energy is turned into heat energy. When gas is burned in a car's engine, its chemical energy is turned into heat energy.
3. **Mechanical energy** is the sum of the stored energy, also called the object's potential energy and the energy of motion also called the kinetic energy of the object, by the virtue of its position. The energy is stored in objects by tension. For example, the compressed springs have mechanical energy, and a stretched rubber has mechanical energy stored.
4. **Nuclear energy** is energy stored in an atom's nucleus or core. The power that keeps the nucleus together is called "strong force." When the nuclei come together or split apart, they can release a lot of energy. For example, when atoms split apart during nuclear fission, they release a lot of energy. The energy can then be used to make electricity. The process of nuclear fission and making electricity is controlled by a set of machines called a nuclear power plant.
5. **Gravitational energy** is the energy stored in an object's height above the earth. More gravitational potential energy is stored in an object when it is higher up. The more energy an object has, the more weight it has. When water falls from a great height and goes through a hydroelectric turbine, its gravitational energy is turned into electricity. In the same way, when someone rides a bike downhill, the bike speeds up because the gravitational energy is turned into motion energy.
6. **Kinetic energy** or the energy of motion is the energy possessed by an object, by the virtue of its motion. Waves, electrons, atoms, molecules, substances, and things may all be in motion. The potential energy of a body at rest is transformed into kinetic energy the moment the body is placed in motion.
7. **Radiant energy** is the energy of electromagnetic and gravitational radiation, that travels in transverse waves. Visible light, X-rays, thermal radiation, gamma rays, and radio waves are all examples of radiative energy. Sunlight is a source of the light and radiant energy needed to power machines and provide the heat essential to supporting life.
8. **Thermal energy** or heat energy is the energy that comes from the faster movement of atoms and molecules in a substance when heat is provided to them. When these particles are in motion, they generate more heat. As its name implies, geothermal energy refers to the earth's inherent thermal energy (geo = earth, thermal = heat).
9. **Motion energy** is energy stored in the movement of objects. More energy is stored in anything that is moving quickly. A reduction in velocity results in the release of the energy needed to keep the item in motion. The wind is a good

illustration of kinetic energy. It happens, for instance, when a vehicle stops suddenly and all of its stored momentum is released at once, resulting in an accident.

10. **Sound energy** is physical vibration or wave that travels through the air by the movement of air molecules, that reaches the ears, and the brain is signalled to interpret the vibration wave as an incoming sound. When an item or material is subjected to a force that causes it to vibrate, noise is created. A longitudinal (compression/rarefaction) wave conducts the energy from one end of the material to the other.
11. **Electrical energy** is the power sent over a wire by electrons, which are charged particles. Electrical energy may be seen in nature during thunderstorms, for instance, as shown by lightning.

1.4 Energy Resources

The advancement of civilization depends on a steady supply of energy. The rising global population can now enjoy longer, healthier lives as a result of economic growth and increased access to energy. There are many different kinds of primary energy sources, such as nuclear energy, fossil energy like oil, coal, and natural gas, and renewable energy sources like wind, solar, geothermal, and hydropower. These primary sources are turned into electricity, which is a secondary source of energy. Electricity flows to your home or business through power lines and other infrastructure. Renewable source of energy refers to the energy derived from the renewable source, and it is also stated as "clean energy" or "green energy". It has gained a lot of popularity as it is an innovation which brings down energy costs and promises a cleaner, greener future, as the amount of pollution generated by the use of renewable sources is lesser than the amount of pollution non-renewable sources creates (Zhu and Wu 2021). It is obtained from a natural source which is replenished constantly within a shorter period. In the power industry, renewable energy sources are gradually but certainly taking the place of "dirty" energy sources such as fossil fuels, such as coal and oil, since renewable energy sources provide the advantage of reduced levels of emissions and pollution (Ashrafi Goudarzi et al. 2019; Moser-Reischl et al. 2018). They are a dependable source of power supply and fuel diversity in addition to being a rapidly expanding source of clean energy with minimal or no carbon emissions (Gago et al. 2013; Moutinho and Madaleno 2021; Pham et al. 2018). These factors lead to increased energy security, protection of the environment, and preservation of natural resources and ecosystems. They also reduce the need for costly fuel imports and accidents such as the Deepwater Horizon oil spill, which occurred in 2010, and resulted in the deaths of 11 employees on the rig as well as millions of marine animals, including mammals, birds, and fish (Gur et al. 2020; Nambi et al. 2017; Oyewole 2018). Recently there was a gas leak due to a pipeline bursting under the sea, causing a bright flame to appear coming from underwater, described as an "eye of fire" up to the Gulf of Mexico's surface (Bartorila and Diaz Montemayor 2021; Wong et al. 2019). The mishap took place on 2 July 2021, on the west of Mexico's Yucatán Peninsula.

(a) Solar energy—is a type of renewable energy, as the sun shines every next day. Humans have been harnessing sunlight from always to survive. Solar cells or photovoltaic cells are used to trap the sun's energy which is then converted into electricity.
(b) Wind energy—earlier windmills were used to grind grains and do other jobs, but now turbines are used to harness the wind energy, wind turns the blades of the turbine, and the electric generator produces electricity.
(c) Hydroelectric power is about harnessing the kinetic energy of the flowing water. When the rapidly descending water falls from a high point, spinning the generator's turbine blades, the energy of the force of water is converted into electricity.
(d) Biomass energy—when the organic matter is burned, the chemical energy in the biomass is converted to thermal energy. It is often thought of as a cleaner source, but it is not. Though biomass energy is a renewable source of energy, but burning biomass leads to more carbon emissions than fossil fuels.
(e) Geothermal energy—The radioactive particles in the core of the earth slowly decay to release a large amount of heat. Geothermal plants are used to harness this heat from the core of the earth.
(f) The tidal energy of the waves is based on dam-like structures or ocean floors, anchored devices that are located on or near the water's surface. But tidal barrages are harmful to marine life.

A non-renewable resource is located under the earth's surface. It is a naturally occurring material which depletes at a rate faster than it is renewed (Chapman et al. 2015; Da Silva Pinto Vieira et al. 2018). Therefore, the consumption rate is very high, and if this rate continues, we will run out of these resources over time. So, we can say that there is a scarcity of non-renewable (Dovì et al. 2009; Waheed et al. 2019). They contain a restricted number of plants, algae, and plankton that flourished in these ancient wetlands. These wetlands are quite old. They took in the light of the sun and generated energy via a process called photosynthesis. After passing away, the creatures sank to the bottom of the lake or ocean where they had been living. When the plants and animals passed away, they left behind a pool of stored energy (Du et al. 2022; Wang et al. 2017). Over time, the decomposing vegetation was ground up by the seafloor. They are then buried beneath an accumulation of rocks and other material, which causes the subterranean environment to become very hot and pressurized (Raza et al. 2013). The plant and animal remain that are left behind in this ecosystem will ultimately be converted into fossil fuels. Fossil fuels such as coal, natural gas, oil, and other substances are examples of non-renewable resources that are used in the production of electricity (Agostinho et al. 2021; Kumar 2020; Misra and Kumar 2020). Nuclear energy is an example of another resource that cannot be replenished.

(a) Oil—this non-renewable resource is drawn out in the liquid form. Crude oil or petroleum oil is used extensively for transportation purposes. We extract crude oil by drilling vertically into the earth's surface. It can be discovered between the layers of the earth's crust. Crude oil is formed when centuries ago tiny sea organisms were buried under the sediments and over time they were decayed and transformed into petroleum. Crude oil is further refined and used to make other products like gasoline and diesel. It is also used in the production of plastics and many other products. This widespread use has resulted in the depletion of the oil reserves and an increase in the level of pollution (An et al. 2016; Sasaki and Sakamoto 2005).

(b) Coal—is extracted from the earth's surface in solid form. Coal is formed when dead plants and animals are buried inside the earth and with time they perish the heat and pressure compress them and result in the accumulation of hydrocarbon which leads to the formation of coal. The highest levels of carbon are found in coal. It emits the highest amount of greenhouse gases when it is burnt (Ayoobi et al. 2021). Coal reserves have been depleting at a very fast rate due to their increased usage for various purposes like industries, production of electricity, etc. (Azapagic 2004; Li et al. 2021a, b).

(c) Natural gas –is extracted from the earth's surface in gaseous form. It is made from the remnants of little marine plants and animals that were buried under the sediments, millions of years ago (Ashrafi Goudarzi et al. 2019; Tanskanen et al. 2021). Due to the high temperature and high pressure, these remains turn into natural gas. It is composed of many gases, but the primary component is methane gas and other gases include propane, ethane, and butane. It has various uses such as it is used for cooking, etc.

(d) Nuclear energy—Atomic nuclei are the source of nuclear energy. Energy is produced either through nuclear fusion or nuclear fission (Ayağ and Samanlioglu 2020; Sim 2020). Radioactive substances, the most common are uranium which occurs naturally and is mined and processed to produce nuclear energy (Agyekum et al. 2020).

1.5 Energy Efficiency

The term "energy efficiency" refers to the practice of utilizing less energy to accomplish the same goal. The heating, cooling, and powering of energy-efficient homes and buildings is reduced, as is the amount of energy used in the production of goods in energy-efficient factories. One of the simplest and most cost-effective strategies to fight global warming, lower energy prices for consumers, and boost competitiveness is to increase energy efficiency. When aiming for carbon neutrality through decarbonization, energy efficiency is equally crucial. When energy is transformed from one form to another, part of the energy that was previously accessible is squandered, and the amount of output of energy produced in proportion to the amount of energy which was initially used as the input for conversion is energy efficiency, is a significant conversion related factor.

$$\eta = \mathrm{E_{out}} / \boldsymbol{E}_{\mathbf{in}} * \mathbf{100\%}$$

- η is the efficiency (expressed as a percentage),
- $\boldsymbol{E}_{\mathbf{out}}$ is the energy output (in Joules), and
- $\boldsymbol{E}_{\mathbf{in}}$ is the energy input (also in Joules).

A rate of efficiency of 0% means that none of the input energy is retained and that there is no production of usable energy (French et al. 2019; Kennedy et al. 2007; Supe et al. 2020). On the other side, efficiency of one hundred per cent shows that there is absolutely no loss of energy at all (Su et al. 2021; Tallman et al. 2021; Xiu et al. 2020). According to the first rule on the conservation of energy, there is no such thing as the creation of new energy. Therefore, the effectiveness of any machine can never reach levels higher than 100%. The ratio of the amount of energy that is produced to the amount of energy that is entered is the definition of energy efficiency (Alzaza et al. 2021; Wenzel et al. 2021). When energy or heat is delivered to a machine for conversion, such as an automobile engine, a certain amount of energy is lost, and only a portion of the total quantity is transformed into the work that the machine is capable of producing. Therefore, we can say that the machine is more efficient if the amount of wasted energy is lower; moreover, the greater the output, the more efficient the machine is. Nicolas Sadi Carnot, a scientist and engineer, devised an ideal engine that optimizes the production of energy in the nineteenth century. This engine, known as the Carnot engine, is a perfect heat engine and thermodynamic cycle (Ramakreshnan et al. 2019; Zandi et al. 2021). Carnot is credited with discovering this engine (called the Carnot cycle). The Carnot cycle is the most efficient theoretical thermodynamic cycle there is for collecting work from two different sources that remain at a constant temperature (Mohamed et al. 2021; Wang et al. 2020). It results in a particular efficiency that is known as the Carnot efficiency. He also noted that it is impossible for any machine, not even a flawless one, to have an efficiency of 100%; it will always be lower than 100%.

1.6 Energy Conversion

The way the globe consumes energy is influenced by policy, technology, and consumer choices. Both motorists have an impact on one another. The interplay shifts and adapts over time and in response to local conditions (such as resources and public support). To better understand how the world's energy system is interconnected, we are constantly investigating energy demand and creating models that evaluate its potential impact. In today's technological world, we can accomplish more with less effort. Most of the times, government policies and commercial frameworks help propel the most promising innovations to widespread adoption. After being sparked into action by policies like tax incentives, recent technologies must compete in the free market to grow to a size where they can have an impact on international markets. A "pull effect," generated by consumers' inclinations, can also drive up the market's desire for innovations. Planning for the future of a company requires an in-depth familiarity with the principles of long-term energy. Energy supply and demand trends, the magnitude and diversity of global energy requirements, the availability, viability, and affordability of alternative energy sources, such as those with lower emissions, the technology and policy frameworks for reducing greenhouse gas emissions, and others are essentials. The Outlook takes these basics into account as the foundation for strategic business and investment decisions and research initiatives over the long term. This forecast considers the expected growth in energy consumption along with the expected growth in energy supply around the world until the year 2050. It considers future and present developments in technology, public policy, consumer tastes, international politics, and economic growth.The definition of energy efficiency is the reduction of energy use while maintaining the same load on the grid. Investing in more efficient electrical appliances and gadgets; for instance, switching from incandescent to compact fluorescent or light emitting diode (LED) bulbs. To save energy, we should do things like switch off the lights when we leave a room, dry our clothes and dishes by air, and use the appropriate-sized burner while cooking. Instead of using the elevator, opting to take the stairs saves energy since the elevator will run less often, but still requires the same amount of power to run. It is more efficient to have more than two persons in the elevator at once than just one. If energy service demand is not decreasing then increasing the energy efficiency will reduce the energy consumption, but not always. High appliance efficiency standards with decreased energy consumption and cost, and consumers can use that money to buy a larger appliance, and this is called the rebound effect. Energy management and efficiency improvements have long been vital for business and commerce (Chen et al. 2018; McNeely 2021). There was a noticeable uptick in attention to energy efficiency in policy circles around the world. Energy efficiency's significance in reaching climate objectives, as well as the size of the economic potential it provides, is becoming more widely acknowledged (Bezyk et al. 2021; Elmqvist et al. 2018; Well and Ludwig 2021). Efficiency is now the first fuel, as per the IEA, although it was the fifth fuel in the 1980s. Advantages outside of the energy sector, such as increased sales, better health and well-being, and macro-benefits like employment growth, have been there for a while but are just now being fully recognized (Dheeraja et al. 2022). Energy efficiency experts should have a deeper comprehension of the non-energy benefits and the potential added value they may offer to an investment (Karaev et al. 2020; Martinez 2020; Rena et al. 2022). Things like higher sales or improved employee and customer health and well-being are considerably more crucial to companies than merely energy savings (Zucaro et al. 2014). As a result, the return on energy savings alone is considerably less likely to get a project authorized. Institutional investors are becoming more committed to energy efficiency, even if most of the pledges have yet to be implemented. Through the Investor Confidence Project, the essential infrastructure for standardization has been established. In reaction to the oil crisis of the 1970s, new ways of thinking about energy began developing.

1.7 Sustainable Development Goals (SDGs)

In 2012, during the United Nations Conference on Sustainable Development in Rio de Janeiro, the Sustainable Development Goals were established (Kwiatkowski et al. 2020; Ravanelli et al. 2018). These objectives were an attempt to create a worldwide agreed-upon strategy to address the world's most pressing environmental, political, and economic problems. In 2015, 193 countries voted to embrace the Sustainable

Development Goals (Galodha and Gupta 2021; Hallström et al. 2015; Hysa 2021). It is related to affordable and clean energy. It is to ensure access to affordable, reliable, modern, and sustainable energy. Making renewable energy more reliable, easy to availability, cheaper, and more efficient. As fossil fuels are unsustainable and harmful to the planet. Some of the indicators of the 7th SGD are:

- Indicator 7.2.1 is renewable energy share in the total final energy consumption. Ensure that renewable energy sources account for a growing percentage of the world's energy demand by 2030. Solar, wind, geothermal, hydropower, bioenergy, and marine renewable energy as a share of total energy use (not primary energy). The energy mix includes sources for transportation, home heating, and cooking.
- Indicator 7.A.1 helps poor nations to fund clean energy R&D and renewable energy generation, particularly hybrid systems. Promote investment in energy infrastructure and clean energy technology and increase international collaboration to increase access to clean energy research and technology by the year 2030. This includes renewable energy, energy efficiency, and advanced and cleaner fossil-fuel technologies.
- Indicator 7.B.1 talks about the ratio of energy efficiency spending to GDP, as well as the amount of FDI used to fund the transfer of infrastructure and technology for sustainable development services. Improve existing systems and introduce cutting-edge innovations to ensure that all residents of developing nations have access to modern, sustainable energy services by the year 2030.
- Indicator 7.3.1 is energy consumption as a percentage of gross domestic product. Increase energy efficiency twice as fast as the world already does by the year 2030. A gauge for this is the energy intensity of national economies (collectively across all sectors). When calculating energy intensity, the number of kilowatt-hours produced per international dollar of GDP in 2011 is considered. This is a measure of how effectively energy is used.

1.8 Impact on Energy Economics

Primary energy intensity is measured as the ratio of energy supply to GDP when PPP is used. Energy intensity refers to the quantity of energy required to produce one dollar's worth of economic activity. Reduced input-to-output ratios indicate more efficient use of energy. The world's energy intensity, measured as GDP per unit of energy used, decreased by around a third between 1990 and 2015. The amount of energy required to do daily tasks has decreased in nearly every part of the world. Differences in energy intensity across countries and regions may be attributed to several underlying factors, including economic structure, climate, and geography (Bassuk et al. 2015; Brink et al. 2016; Xie and Schöbel-Rutschmann 2015). Manufacturing-centric economies consume more energy per dollar of GDP than service-centric countries. More energy is needed to heat and cool countries and regions with greater temperature changes. The distance between cities and the infrastructure found inside them both affect the amount of energy required to convey goods and people. Different regions have different laws on energy efficiency, and the relative efficiency of buildings, vehicles, and industries is heavily influenced by local rules, incentives, and market competition (Bouadila et al. 2014; Tian et al. 2017). The globe has become more energy efficient, according to statistics from the U.S. Energy Information Administration (EIA), as the ratio of energy supply to GDP when measured at purchasing power parity has exhibited a declining trend (Kurlovich et al. 2021).

1.9 Energy Intensity and Sustainable Development Scenario

In 2019, the global economy used 2% less energy per unit of GDP than it did in 2018. This rise in energy intensity balanced for around 70% of the increase in global energy consumption that would have resulted from economic development (Horani 2020; Luthra et al. 2017; McNeely 2021; Moktadir et al. 2018). This shift can be ascribed to advances in energy efficiency, which account for about 70% of the reduction in energy consumption. More mild weather in major regions contributed more than 20%, mostly in the form of lower heating and cooling demand. After adjusting for weather, the global economy's energy intensity improved by 1.6% in 2019. Additionally, switching fuels resulted in a 10% increase in energy efficiency. Changes in power generating composition in China, Europe, and the USA substituted thermal power generation with a mix of renewables, efficient natural gas, and nuclear, reducing energy losses in the combustion and conversion processes (Chow and Bakar 2019; de Manuel et al. 2021; Kim and Choi 2022; Niemi et al. 2012). Changes in the composition of the world economy hindered the pace of improvement in energy intensity throughout the world in 2019. Although energy efficiency was the most important factor in improving energy intensity in 2019, overall energy savings from efficiency were 11% lower than in 2018. Because policy coverage has increased continuously but only little in recent years, the majority of efficiency improvements are attributable to the replacement of equipment covered by existing policies rather than new policies (Amudha and Murali 2021; Ishitsuka et al. 2021; Ji et al. 2021; Obrist et al. 2021). Increasing the breadth and robustness of norms and standards is critical if energy intensity is to be improved quickly enough to meet the SDS. In 2019, the necessary yearly rate of over 3% is roughly double the actual pace of structural improvement.

1.10 Benefits of Solar Energy Utilisation

(a) Given that it may be replenished over time, the advantages are many. Here are a few examples:

1. If you own a home, business, or other commercial property, switching to solar power might eliminate or greatly decrease your monthly electricity costs. Since solar panels may be improved upon and last for decades (about 20–25), their use can result in drastically reduced or eliminated electrical costs.

2. Using solar power significantly decreases household carbon emissions. The given energy is clean and renewable; thus, its environmental impact is minimal.

(b) By switching to solar power, you can stop worrying about ever-increasing electricity costs. Large families and the usage of power-hungry gadgets contribute to the high energy costs faced by many families (appliances like geysers and Air Conditioners). In light of this, putting in solar panels and utilizing the energy to power some of the appliances may drastically cut prices and give you some control over the ever-increasing energy bills.

1.11 Limitations of Solar Energy

1. The local climate has a significant role in determining the viability of solar energy collection. It was already said in the introduction that areas with significant precipitation and humidity are less likely to capture as much power as dry ones. So, solar energy may be effectively collected, if the right conditions exist.
2. The kind of roof home or business has might also affect how easy it is to set up solar panels. Roof decks and roof tiles might cause complications when installing solar panels. Even in large apartments and flats, some individuals choose to install skylights, which might lead to issues down the road. Attaching a mounting system, also known as racking, is used to set up solar panels. Solar panel installation might be difficult on slate roofs for the same reason.
3. The kind of roof home or business has might also affect how easy it is to set up solar panels. Roof decks and roof tiles might cause complications when installing solar panels. Even in large apartments and flats, some individuals choose to install skylights, which might lead to issues down the road. Attaching a mounting system, also known as racking, is used to set up solar panels. Solar panel installation might be difficult on slate roofs for the same reason.

1.12 Efficiency and Management Scenarios

Saving energy is crucial in the battle against climate change because conventional energy sources burn fossil fuels, as major sources of greenhouse gases and air pollution. Buildings and households which are energy efficient are also better prepared to use renewable energy sources, which do not emit harmful emissions. Cleaner air, water, and land are immediate benefits of reducing fossil fuel consumption, especially for the most vulnerable populations and those with preexisting ailments made worse by pollution. Using solar panels is good for humanity since it may lessen the need to generate power from polluting sources. Energy collected in this way is renewable as long as the sun shines. There are benefits and drawbacks to using solar energy, but it is becoming more common in developed countries, with residential and commercial buildings all over the world installing solar panels and even whole solar farms. Especially in rural regions, where blackouts are widespread, the Indian government promotes the use of solar panels. Solar-powered streetlights are already commonplace, particularly in India's major cities. While solar is the most practical kind of renewable energy in terms of installation and harnessing, there are numerous more options. In the not-too-distant future, more sophisticated semiconductors will be developed, which may boost the electrical energy production of photovoltaic cells or solar cells.

With the right use of the chosen methodologies, this case study may be able to provide light on the difficulties and research questions given at the outset. A new development scenario for an urban neighbourhood may involve a debate on how to make the most effective use of easily accessible local energy sources to create an ideal energy system that can fulfil predicted future energy demands. The ideal solutions for energy systems in these settings account for finite resources and the unpredictability of market pricing. To determine the ecological effect of each energy system choice, the scenario settlement configurations were analysed. It is possible to find an energy system solution that has a negligible effect on the environment while maximizing financial gain.

Managing and coordinating the wide variety of supply and conversion technologies and demand applications is a significant challenge within the smart city setting, underscoring the need of adopting an intelligent strategy. Many of the technological difficulties in integrating energy systems may be overcome by the widespread proliferation of sensors and the installation of embedded computational intelligence algorithms. But there are still obstacles to overcome since appropriate approaches are required to manage large networks of actors, frequently with opposing aims, while making design and operational choices for systems spanning a broad range of characteristics and timelines. While noting some of the fundamental difficulties now encountered and proposing potential routes for the industry, this study looks at the latest innovations in the smart energy sector, focusing on methodologies in the main application areas along with relevant implemented examples. The improvement in energy efficiency includes various design mechanisms and control hierarchies that may be necessary for creation of a optimal framework. The work presents the ability of computational intelligence for smart energy management to showcase the potential solutions for the existing challenges of the domain.

References

Adelaja S, Shaw J, Beyea W, Charles McKeown JD, Agugiaro G, Nex F, Remondino F, Filippi R De, Droghetti S, Furlanello C, Foundation BK, Stevanato G, Filippi R De, Furlanello C, Kessler FB, Modelling U, Unit E, Al-enezi FQ, Sykulski JK (2011). Potential and prospects of solar energy in Uttara Kannada, District of Karnataka State, India. Solar Energy 02(2):1986–2000. https://doi.org/10.1080/00908319708908903. www.imagingnotes.com

Agostinho F, Costa M, Coscieme L, Almeida CMVB, Giannetti BF (2021) Assessing cities growth-degrowth pulsing by emergy and fractals: a methodological proposal. Cities 113. https://doi.org/10.1016/j.cities.2021.103162

Agyekum EB, Ansah MNS, Afornu KB (2020) Nuclear energy for sustainable development: SWOT analysis on Ghana's nuclear agenda. Energy Rep 6. 107–115. https://doi.org/10.1016/j.egyr.2019.11.163

Alzaza A, Ohenoja K, Illikainen M (2021) Enhancing the mechanical and durability properties of subzero-cured one-part alkali-activated blast furnace slag mortar by using submicron metallurgical residue as an additive. Cem Concr Compos 122. https://doi.org/10.1016/j.cemconcomp.2021.104128

Amudha S, Murali M (2021) Implementation of energy efficient fog based health monitoring and emergency admission prediction system using IoT. Webology 18(SpecialIss):171–189. https://doi.org/10.14704/WEB/V18SI02/WEB18065

An D, Xi B, Wang Y, Xu D, Tang J, Dong L, Ren J, Pang C (2016) A sustainability assessment methodology for prioritizing the technologies of groundwater contamination remediation. J Clean Prod 112:4647–4656. https://doi.org/10.1016/j.jclepro.2015.08.020

Ashrafi Goudarzi S, Fazelpour FB, Gharehpetian G, Rosen MA (2019) Techno-economic assessment of hybrid renewable resources for a residential building in tehran. Environmental Progress and Sustainable Energy 38(5):1–14. https://doi.org/10.1002/ep.13209

Ayağ Z, Samanlioglu F (2020) Fuzzy AHP-GRA approach to evaluating energy sources: a case of Turkey. Int J Energy Sect Manage 14(1):40–58. https://doi.org/10.1108/IJESM-09-2018-0012

Ayoobi AW, Ahmadi H, Inceoglu M, Pekkan E (2021) Seasonal impacts of buildings' energy consumption on the variation and spatial distribution of air pollutant over Kabul City: application of Sentinel—5P TROPOMI products. Air Qual Atmos Health. https://doi.org/10.1007/s11869-021-01085-9

Azapagic A (2004) Developing a framework for sustainable development indicators for the mining and minerals industry. J Clean Prod 12(6):639–662. https://doi.org/10.1016/S0959-6526(03)00075-1

Aziz AS, Tajuddin MFNB, Adzman MRB (2018) Feasibility analysis of PV/Wind/Battery hybrid power generation: a case study. Int J Renew Energy Res 8(2):661–671

Balasubramanian B, Ariffin AM (2014) A review on feasibility assessment of solar photovoltaic systems 4(7):192–201

Bartorila MA, Diaz Montemayor G (2021) Urban and environmental regeneration in Mexican cities: a design framework. J Urban. https://doi.org/10.1080/17549175.2021.1893796

Basir R, Aziz N, Ahmad SS, Wahid A (2013) Satellite remote sensing for identification of solar potential sites in Pakistan. Int J Basic Appl Sci 2(2):200–209

Bassuk NL, Universite AB, Jean M, Universite C, Theoretical L, Politics U, Book TC, Planning T, Lemon JT, Olorunfemi JF, Harloe M, Helm C, Germany U, Technologies G, Juppenlatz M, Dale PF, Way W, Lambin EF, Strahler AH, Bibliography AA (2015) On using landscape metrics for landscape similarity search. Landscape and Urban Plann 117(1):1–12. https://doi.org/10.1038/srep11160

Bataineh K (2012) Optimal configuration for design of stand-alone PV system. Smart Grid and Renew Energy 03(02):139–147. https://doi.org/10.4236/sgre.2012.32020

Bellezoni RA, Meng F, He P, Seto KC (2021) Understanding and conceptualizing how urban green and blue infrastructure affects the food, water, and energy nexus: a synthesis of the literature. J Cleaner Prod 289. https://doi.org/10.1016/j.jclepro.2021.125825

Bezyk Y, Sówka I, Górka M (2021) Assessment of urban CO2 budget: anthropogenic and biogenic inputs. Urban Clim 39. https://doi.org/10.1016/j.uclim.2021.100949

Bian J, Li A, Lei G, Zhang Z, Nan X (2020) Global high-resolution mountain green cover index mapping based on Landsat images and Google Earth Engine. ISPRS J Photogramm Remote Sens 162:63–76. https://doi.org/10.1016/j.isprsjprs.2020.02.011

Borozan D, Pekanov Starcevic D (2018) Unveiling the factors affecting final electricity consumption: does the regional component matter? Int J Energy Sect Manage 12(2):202–220. https://doi.org/10.1108/IJESM-10-2017-0011

Bouadila S, Lazaar M, Skouri S, Kooli S, Farhat A (2014) Energy and exergy analysis of a new solar air heater with latent storage energy. Int J Hydrogen Energy 39(27):15266–15274. https://doi.org/10.1016/j.ijhydene.2014.04.074

Brink E, Aalders T, Ádám D, Feller R, Henselek Y, Hoffmann A, Ibe K, Matthey-Doret A, Meyer M, Negrut NL, Rau A-L, Riewerts B, von Schuckmann L, Törnros S, von Wehrden H, Abson DJ, Wamsler C (2016) Cascades of green: a review of ecosystem-based adaptation in urban areas. Glob Environ Chang 36:111–123. https://doi.org/10.1016/j.gloenvcha.2015.11.003

Chapman B, McDonald K, Shimada M, Rosenqvist A, Schroeder R, Hess L (2015) Mapping regional inundation with spaceborne L-Band SAR. Remote Sens 7(5):5440–5470. https://doi.org/10.3390/rs70505440

Chen C-W, Wang J-H, Wang JC, Shen Z-H (2018) Developing indicators for sustainable campuses in Taiwan using fuzzy Delphi method and analytic hierarchy process. J Clean Prod 193:661–671. https://doi.org/10.1016/j.jclepro.2018.05.082

Chen SP, Zeng S, Xie CG (2000) Remote sensing and GIS for urban growth analysis in China. Photogramm Eng Remote Sens 66(5):593–598. https://doi.org/10.1007/978-94-007-4698-5

Chow MF, Bakar MFA (2019) Environmental benefits of green roof to the sustainable urban development: a review. Lect Notes Civ Eng 9:1525–1541. https://doi.org/10.1007/978-981-10-8016-6_110

Colmenar-Santos A, Monteagudo-Mencucci M, Rosales-Asensio E, de Simón-Martín M, Pérez-Molina C (2019) Optimized design method for storage systems in photovoltaic plants with delivery limitation. Sol Energy 180(February 2018) 468–488. https://doi.org/10.1016/j.solener.2019.01.046

Da Silva Pinto Vieira RM, Do Amaral Cunha APM, Barbosa AA, Neto GGR, Tomasella J, Dos Santos Alvalá RC, Santos FC, De Oliveira Santana M (2018) Analysis of the interactions between climate data and the desertification process in the desertification hotspot of Cabrobó- PE, Brazil. [Análise das interações entre dados climáticos e o processo de desertificação no núcleo de desertificação de Cabrobó-PE. Sustentabilidade Em Debate 9(2):72–87. https://doi.org/10.18472/SUSTDEB.V9N2.2018.8542

de Manuel BF, Méndez-Fernández L, Peña L, Ametzaga-Arregi I (2021) A new indicator of the effectiveness of urban green infrastructure based on ecosystem services assessment. Basic Appl Ecol 53:12–25. https://doi.org/10.1016/j.baae.2021.02.012

Dheeraja K, Padma Priya R, Ritika T (2022) Optimal real-time pricing and sustainable load scheduling model for smart homes using stackelberg game theory. In: Lecture notes on data engineering and communications technologies, Vol 99. https://doi.org/10.1007/978-981-16-7182-1_22

Dover JW (2018) Introduction to urban sustainability issues: Urban ecosystem. Elsevier Inc., In Nature Based Strategies for Urban and Building Sustainability. https://doi.org/10.1016/B978-0-12-812150-4.00001-X

Dovì VG, Friedler F, Huisingh D, Klemeš JJ (2009) Cleaner energy for sustainable future. J Clean Prod. https://doi.org/10.1016/j.jclepro.2009.02.001

Du S, Zhang X, Jin X, Zhou X, Shi X (2022) A review of multi-scale modelling, assessment, and improvement methods of the urban thermal and wind environment. Build Environ 213. https://doi.org/10.1016/j.buildenv.2022.108860

Elmqvist T, Siri J, Andersson E, Anderson P, Bai X, Das PK, Gatere T, Gonzalez A, Goodness J, Handel SN, Hermansson Török E, Kavonic J, Kronenberg J, Lindgren E, Maddox D, Maher R, Mbow C, McPhearson T, Mulligan J, Vogel C (2018) Urban tinkering. Sustain Sci 13(6):1549–1564. https://doi.org/10.1007/s11625-018-0611-0

French A, Mozaffari M, Eldosouky A, Saad W (2019) Environment-aware deployment of wireless drones base stations with Google Earth Simulator. In: 2019 IEEE international conference on pervasive computing and communications workshops, PerCom workshops, pp 868–873. https://doi.org/10.1109/PERCOMW.2019.8730845

Gago EJ, Roldan J, Pacheco-Torres R, Ordóñez J (2013) The city and urban heat islands: a review of strategies to mitigate adverse effects 25:749–758. https://doi.org/10.1016/j.rser.2013.05.057

Galodha A, Gupta SK (2021) Land surface temperature as an indicator of urban heat island effect: a google earth engine based Web-App. In: A. M, Schuckman K (ed), International archives of the photogrammetry, remote sensing and spatial information sciences—ISPRS Archives, vol 44, no M–3, pp 57–62). International society for photogrammetry and remote sensing. https://doi.org/10.5194/isprs-archives-XLIV-M-3-2021-57-2021

Gambhir A, Dixit S, Toro V, Singh V, Ramachandra TV, Jain R, Krishnadas G, Nayak JK, Prajapati JA, Noel, Cressie, Christopher, KW, Lean JL, Sayigh AAM, Pareek A, Studios B, Commission CE, National T, Plan A, Change C, Nehru J, Observatory N (2010) A guide to PV system design and installation. In: Solar energy, vol 44, no 5. Academic Press. https://doi.org/10.1002/wcc.018

Gur FA, Bendickson JS, Madden L, McDowell WC (2020) Entrepreneurial opportunity recognition in the face of disasters. Int J Entrepreneurial Behav Res 26(4):671–693. https://doi.org/10.1108/IJEBR-09-2019-0537

Hallström E, Carlsson-Kanyama A, Börjesson P (2015) environmental impact of dietary change: a systematic review. J Cleaner Prod. https://doi.org/10.1016/j.jclepro.2014.12.008

Hardiman BS, Wang JA, Hutyra LR, Gately CK, Getson JM, Friedl MA (2017) Accounting for urban biogenic fluxes in regional carbon budgets. Sci Total Environ 592:366–372. https://doi.org/10.1016/j.scitotenv.2017.03.028

He Y, Liu T, Wang T, Liang X, Wei H, Zheng Z, Xiao X (2021) Case analysis of integrated maintenance technology for multi-dimensional rapid detection and trenchless reinforcement. In: M. S. Mostafa MMH (ed) E3S Web of Conferences, vol 261, EDP Sciences. https://doi.org/10.1051/e3sconf/202126102087

Horani LF (2020) Identification of target customers for sustainable design. J Cleaner Prod 274. https://doi.org/10.1016/j.jclepro.2020.123102

Hysa A (2021) Introducing transversal connectivity index (TCI) as a method to evaluate the effectiveness of the blue-green infrastructure at metropolitan scale. Ecol Indic 124. https://doi.org/10.1016/j.ecolind.2021.107432

Iqbal A, Iqbal T (2019) Design and analysis of a stand-alone PV system for a rural house in Pakistan. Int J Photoenergy 8

Ishitsuka Y, Gleason CJ, Hagemann MW, Beighley E, Allen GH, Feng D, Lin P, Pan M, Andreadis K, Pavelsky TM (2021) Combining optical remote sensing, McFLI discharge estimation, global hydrologic modeling, and data assimilation to improve daily discharge estimates across an entire large watershed. Water Resour Res 57(3). https://doi.org/10.1029/2020WR027794

Ji Y, Kumar R, Singh D, Singh M (2021) Performance analysis of target information recognition system for agricultural robots. Int J Agric Environ Inf Syst 12(2):49–60. https://doi.org/10.4018/IJAEIS.20210401.oa4

Jin X, Zhu X, Xue Y (2019) Satellite-based analysis of regional evapotranspiration trends in a semi-arid area. Int J Remote Sens 40(9):3267–3288. https://doi.org/10.1080/01431161.2018.1500727

Karaev A, Ponkratov V, Masterov A, Pozdnyaev A, Kuznetsov N (2020) The analysis of the comparative efficiency of state support for natural gas production: the case of some OECD and brics countries. Entrepreneurship Sustain Issues 7(3):1778–1789. https://doi.org/10.9770/jesi.2020.7.3(22)

Kaur M, Dhundhara S, Verma YP, Chauhan S (2020) Techno-economic analysis of photovoltaic-biomass-based microgrid system for reliable rural electrification. Int Trans Electr Energy Syst 30(5):1–20. https://doi.org/10.1002/2050-7038.12347

Kennedy C, Cuddihy J, Engel-Yan J (2007) The changing metabolism of cities. J Ind Ecol 11(2):43–59. https://doi.org/10.1162/jie.2007.1107

Khan M, Mehran MT, Haq ZU, Ullah Z, Naqvi SR, Ihsan M, Abbass H (2021) Applications of artificial intelligence in COVID-19 pandemic: a comprehensive review. Expert Syst Appl 185. https://doi.org/10.1016/j.eswa.2021.115695

Khanna A (2012) Smart grid, smart controllers and home energy automation—creating the infrastructure for future. Smart Grid and Renew Energy 03(03):165–174. https://doi.org/10.4236/sgre.2012.33024

Kim, KG, Choi HS (2022) Market mechanisms in the building sector from an urban perspective: creative building solutions. Urban Book Ser 361–388. https://doi.org/10.1007/978-3-030-80165-6_10

Kumar D (2019) Hyper-temporal variability analysis of solar insolation with respect to local seasons. Remote Sens Appl Soc Environ 15:100241. https://doi.org/10.1016/j.rsase.2019.100241

Kumar D (2020) Urban energy system management for enhanced energy potential for upcoming smart cities. Energy Explor Exploit. https://doi.org/10.1177/0144598720937529

Kurlovich DM, Usova IP, Sysoyeva VA (2021) Cartographic and geoinformational support for development of green urban planning in the Republic of Belarus. InterCarto InterGIS 27:151–164. https://doi.org/10.35595/2414-9179-2021-1-27-151-164

Kwiatkowski G, Hjalager A-M, Liburd J, Saabye Simonsen P (2020) Volunteering and collaborative governance innovation in the Wadden Sea National Park. Curr Issue Tour 23(8):971–989. https://doi.org/10.1080/13683500.2019.1571022

Lai CS, Locatelli G (2021) Are energy policies for supporting low-carbon power generation killing energy storage?. J Cleaner Prod 280. https://doi.org/10.1016/j.jclepro.2020.124626

Lee C-Y, Wang K, Sun W (2021) Allocation of emissions permit for China's iron and steel industry in an imperfectly competitive market: a nash equilibrium DEA method. IEEE Trans Eng Manage 68(2):548–561. https://doi.org/10.1109/TEM.2019.2904985

Lemlin J, Cloughley M, Oldham R (1997) Facing the future with confidence-a paper for the world petroleum congress of 1997-a joint IPIECA/E and P forum/OCIMF contribution. In: World Petroleum Congress Proceedings, 1997-Oct, pp 251–259. https://www.scopus.com/inward/record.uri?eid=2-s2.0-85057877239&partnerID=40&md5=b50c9e2b873ca5a022d4d5e573bab7da

Leslie GW, Stern DI, Shanker A, Hogan MT (2020) Designing electricity markets for high penetrations of zero or low marginal cost intermittent energy sources. Electricity J 33(9). https://doi.org/10.1016/j.tej.2020.106847

Li J, Yan X, Yan X, Guo W, Wang K, Qiao J (2021a) Temporal and spatial variation characteristic of vegetation coverage in the yellow river basin based on GEE cloud platform. Meitan Xuebao/J Chin Coal Soc 46(5):1439–1450. https://doi.org/10.13225/j.cnki.jccs.ST21.0331

Li W, El-Askary H, Lakshmi V, Piechota T, Struppa D (2020) Earth observation and cloud computing in support of two sustainable development goals for the river Nile watershed countries. Remote Sens 12(9). https://doi.org/10.3390/RS12091391

Li X, Chen WY, Hu FZY, Cho FHT (2021b) Homebuyers' heterogeneous preferences for urban green–blue spaces: a spatial multilevel autoregressive analysis. Landscape and Urban Plann 216. https://doi.org/10.1016/j.landurbplan.2021b.104250

Luthra S, Govindan K, Kannan D, Mangla SK, Garg CP (2017) An integrated framework for sustainable supplier selection and evaluation in supply chains. J Clean Prod 140:1686–1698. https://doi.org/10.1016/j.jclepro.2016.09.078

Marco CD, Ximena HB, Marcelo HB, Rafael PF (2006) Experimental study of the trays dissipator as a level and direction change structure [Estudio experimental del disipador de bandejas como estructura de cambio de nivel y de dirección]. In: International symposium on hydraulic structures—XXII Congreso Latinoamericano de Hidraulica. https://www.scopus.com/inward/record.uri?eid=2-s2.0-85088066863&partnerID=40&md5=8999d76ee6f3d829889ada3a5e8f4609

Martinez N (2020) Resisting renewables: The energy epistemics of social opposition in Mexico. Energy Res Soc Sci 70(101632). https://doi.org/10.1016/j.erss.2020.101632

McNeely JA (2021) Nature and COVID-19: The pandemic, the environment, and the way ahead. Ambio 50(4):767–781. https://doi.org/10.1007/s13280-020-01447-0

Misra M, Kumar D (2020) A hybrid indexing approach for sustainable smart cities development. J Indian Soc Remote Sens 48(11):1639–1643. https://doi.org/10.1007/s12524-020-01171-y

Mohamed M, Othman A, Abotalib AZ, Majrashi A (2021) Urban heat island effects on megacities in desert environments using spatial network analysis and remote sensing data: a case study from western Saudi Arabia. Remote Sens 13(10). https://doi.org/10.3390/rs13101941

Moktadir A, Rahman T, Jabbour CJC, Mithun Ali S, Kabir G (2018) Prioritization of drivers of corporate social responsibility in the footwear industry in an emerging economy: A fuzzy AHP approach. J Clean Prod 201:369–381. https://doi.org/10.1016/j.jclepro.2018.07.326

Moser-Reischl A, Uhl E, Rötzer T, Biber P, Van Con T, Tan NT, Pretzsch H (2018) Effects of the urban heat Island and climate change on the growth of Khaya senegalensis in Hanoi, Vietnam. For Ecosyst 5(1). https://doi.org/10.1186/s40663-018-0155-x

Moutinho V, Madaleno M (2021) Assessing eco-efficiency in Asian and African countries using stochastic frontier analysis. Energies 14(4). https://doi.org/10.3390/en14041168

Msheik M, Rodat S, Abanades S (2021) Methane cracking for hydrogen production: a review of catalytic and molten media pyrolysis. Energies 14(11). https://doi.org/10.3390/en14113107

Nambi IM, Rajasekhar B, Loganathan V, RaviKrishna R (2017) An assessment of subsurface contamination of an urban coastal aquifer due to oil spill. Environ Monit Assess 189(4). https://doi.org/10.1007/s10661-017-5833-6

Niemi R, Mikkola J, Lund PD (2012) Urban energy systems with smart multi-carrier energy networks and renewable energy generation. Renew Energy 48:524–536. https://doi.org/10.1016/j.renene.2012.05.017

Obrist MD, Kannan R, Schmidt TJ, Kober T (2021) Decarbonization pathways of the Swiss cement industry towards net zero emissions. J Cleaner Prod 288. https://doi.org/10.1016/j.jclepro.2020.125413

Oral HV, Radinja M, Rizzo A, Kearney K, Andersen TR, Krzeminski P, Buttiglieri G, Ayral-Cinar D, Comas J, Gajewska M, Hartl M, Finger DC, Kazak JK, Mattila H, Vieira P, Piro P, Palermo SA, Turco M, Pirouz B, Carvalho PN (2021) Management of urban waters with nature-based solutions in circular cities—exemplified through seven urban circularity challenges. Water (Switzerland) 13(23). https://doi.org/10.3390/w13233334

Oyewole S (2018) Flying and bombing: the contributions of air power to security and crisis management in the Niger Delta region of Nigeria. Def Stud 18(4):514–537. https://doi.org/10.1080/14702436.2018.1524709

Papaefthymiou G, Dragoon K (2016) Towards 100% renewable energy systems: uncapping power system flexibility. Energy Policy 92:69–82. https://doi.org/10.1016/j.enpol.2016.01.025

Perez R, David M, Hoff TE, Jamaly M, Kivalov S, Kleissl J, Lauret P, Perez M (2016) Spatial and Temporal Variability of Solar Energy. Found Trends® Renew Energy 1(1):1–44. https://doi.org/10.1561/2700000006

Pham TD, Yoshino K, Le NN, Bui DT (2018) Estimating aboveground biomass of a mangrove plantation on the Northern coast of Vietnam using machine learning techniques with an integration of ALOS-2 PALSAR-2 and Sentinel-2A data. Int J Remote Sens 39(22):1–28. https://doi.org/10.1080/01431161.2018.1471544

Pramanick S, Ahmad N, Jain DK (2021) Real time electricity market (RTM) in India. In: 5th International conference on smart grid and smart cities, ICSGSC, pp 112–117. https://doi.org/10.1109/ICSGSC52434.2021.9490431

Ramakreshnan L, Aghamohammadi N, Fong CS, Ghaffarianhoseini A, Wong LP, Sulaiman NM (2019) Empirical study on temporal variations of canopy-level Urban Heat Island effect in the tropical city of Greater Kuala Lumpur. 44:748–762. https://doi.org/10.1016/j.scs.2018.10.039

Ramirez L, Zink R, Dorner W, Stoeglehner G (2015) Spatio-temporal modeling of roof-top photovoltaic panels for improved technical potential assessment and electricity peak load offsetting at the municipal scale. Comput Environ Urban Syst 52:58–69. https://doi.org/10.1016/j.compenvurbsys.2015.03.002

Ravanelli R, Nascetti A, Cirigliano RV Di Rico C, Leuzzi G, Monti P, Crespi M (2018) Monitoring the impact of land cover change on surface urban heat island through Google Earth Engine: Proposal of a global methodology, first applications and problems. Remote Sens 10(9). https://doi.org/10.3390/rs10091488

Raza MQ, Haider MU, Ali SM, Rashid MZ, Sharif F (2013) Demand and response in smart grids for modern power system 133–136

Rena, Yadav S, Patel S, Killedar DJ, Kumar S, Kumar R (2022) Eco-innovations and sustainability in solid waste management: An indian upfront in technological, organizational, start-ups and financial framework. J Environ Manage 302. https://doi.org/10.1016/j.jenvman.2021.113953

Riggs BC, Biedenharn R, Dougher C, Ji YV, Xu Q, Romanin V, Codd DS, Zahler JM, Escarra MD (2017) Techno-economic analysis of hybrid PV/T systems for process heat using electricity to subsidize the cost of heat. Appl Energy 208:1370–1378. https://doi.org/10.1016/j.apenergy.2017.09.018

Rn HK, Shafiullah GM, Maung A, Oo T, Stojcevski A (2014) Feasibility study of standalone hybrid power system modeled with photovoltaic modules and ethanol generator for Victoria. Int J Eng Sci Invention 3(8):54–69

Rowe B (2016) Carbon sequestration and storage: the case for green roofs in urban areas. In sustainable surface water management: a handbook for SUDS. Wiley Blackwell. https://doi.org/10.1002/9781118897690.ch14

Rustemli S, Dincer F (2011) Economic analysis and modeling process of photovoltaic power systems 9:243–247

Saraswat SK, Digalwar A, Yadav SS (2021) Development of assessment model for selection of sustainable energy source in India: hybrid fuzzy MCDM approach. In: Advances in intelligent systems and computing 1197 AISC, pp 649–657. https://doi.org/10.1007/978-3-030-51156-2_75

Sasaki K, Sakamoto K (2005) Vertical differences in the composition of PM10 and PM 2.5 in the urban atmosphere of Osaka, Japan. Atmos Environ 39(38):7240–7250. https://doi.org/10.1016/j.atmosenv.2005.09.004

Shah SAA, Solangi YA, Ikram M (2019) Analysis of barriers to the adoption of cleaner energy technologies in Pakistan using modified delphi and fuzzy analytical hierarchy process. J Clean Prod 235:1037–1050. https://doi.org/10.1016/j.jclepro.2019.07.020

Sim LC (2020) Low-carbon energy in the Gulf: upending the rentier state? Energy Res Soc Sci 70(July):101752. https://doi.org/10.1016/j.erss.2020.101752

Su H, Chi L, Zio E, Li Z, Fan L, Yang Z, Liu Z, Zhang J (2021) An integrated, systematic data-driven supply-demand side management method for smart integrated energy systems. Energy 235. https://doi.org/10.1016/j.energy.2021.121416

Supe H, Avtar R, Singh D, Gupta A, Yunus AP, Dou J, Ravankar AA, Mohan G, Chapagain SK, Sharma V, Singh CK, Tutubalina O, Kharrazi A (2020) Google earth engine for the detection of soiling on photovoltaic solar panels in arid environments. Remote Sens 12(9). https://doi.org/10.3390/RS12091466

Tallman J, Yu D, Pantoja M (2021) Thermal detection for free flight. J Phys Conf Ser 1828(1). https://doi.org/10.1088/1742-6596/1828/1/012133

Tan Z, Lau KKL, Ng E (2016) Urban tree design approaches for mitigating daytime urban heat island effects in a high-density urban environment 114:265–274. https://doi.org/10.1016/j.enbuild.2015.06.031

Tanskanen A, Jantunen A, Saksa JM, Partanen J, Bergman J, Bertsch V, Treitz M, Geldermann J, Rentz O, Di Silvio B, Cesarotti V, Introna V, Babatunde OM, Babatunde DE, Denwigwe IH, Adedoja TB, Adedoja OS, Okharedia TE, Kwakwa PA, Authors F (2021) New directions in UK energy policy. Int J Energy Sect Manage 15(2):157–172. https://doi.org/10.1108/IJESM-12-2019-0001

Tcholtchev N, Schieferdecker I (2021) Sustainable and reliable information and communication technology for resilient smart cities. Smart Cities 4(1):156–176. https://doi.org/10.3390/smartcities4010009

Thomas I, Frankhauser P, Biernacki C (2008) The morphology of built-up landscapes in Wallonia (Belgium): a classification using fractal indices. Landsc Urban Plan 84(2):99–115. https://doi.org/10.1016/j.landurbplan.2007.07.002

Tian G, Zhang H, Feng Y, Jia H, Zhang C, Jiang Z, Li Z, Li P (2017) Operation patterns analysis of automotive components remanufacturing industry development in China. J Clean Prod 164:1363–1375. https://doi.org/10.1016/j.jclepro.2017.07.028

Waheed R, Sarwar S, Wei C (2019) The survey of economic growth, energy consumption and carbon emission. Energy Rep 5:1103–1115. https://doi.org/10.1016/j.egyr.2019.07.006
Wang K, Jiang S, Wang J, Zhou C, Wang X, Lee X (2017) Comparing the diurnal and seasonal variabilities of atmospheric and surface urban heat islands based on the Beijing urban meteorological network 122(4):2131–2154. https://doi.org/10.1002/2016JD025304
Wang Y, Li Y, Xue Y, Martilli A, Shen J, Chan PW (2020) City-scale morphological influence on diurnal urban air temperature 169. https://doi.org/10.1016/j.buildenv.2019.106527
Well F, Ludwig F (2021) Development of an integrated design strategy for blue-green architecture. Sustainability (Switzerland) 13(14). https://doi.org/10.3390/su13147944
Wenzel PM, Radgen P, Westermeyer J (2021) Comparing exergy analysis and physical optimum method regarding an induction furnace. Energies 14(6). https://doi.org/10.3390/en14061621
Wong BA, Thomas C, Halpin P (2019) Automating offshore infrastructure extractions using synthetic aperture radar and Google Earth Engine. Remote Sens Environ 233. https://doi.org/10.1016/j.rse.2019.111412
Xie Y, Schöbel-Rutschmann S (2015) Identifying and reactivating historic cultural landscape elements as green-blue infrastructure for polder urbanization. a case study of Suzhou, Yangtze river delta region. In: History of the future: 52nd world congress of the international federation of landscape architects, IFLA 2015—Congress Proceedings, pp 695–701
Xiu N, Ignatieva M, van den Bosch CK, Zhang S (2020) Applying a socio-ecological green network framework to Xi'an City, China. Landscape Ecol Eng 16(2):135–150. https://doi.org/10.1007/s11355-020-00412-z
Yadav M, Kumar A, Mangla SK, Luthra S, Bamel U, Garza-Reyes JA (2019) Mapping the human resource focused enablers with sustainability viewpoints in Indian power sector. J Clean Prod 210:1311–1323. https://doi.org/10.1016/j.jclepro.2018.11.132
Yamin N, Bhat, G (2021) Online solar energy prediction for energy-harvesting internet of things devices. In: Proceedings of the international symposium on low power electronics and design. https://doi.org/10.1109/ISLPED52811.2021.9502504
Yang J, Ren J, Sun D, Xiao X, Xia J, (Cecilia), Jin C, Li, X (2021) Understanding land surface temperature impact factors based on local climate zones. Sustain Cities Soc 69(March):102818. https://doi.org/10.1016/j.scs.2021.102818
Yaya S, Bishwajit G, Shah V (2016) Wealth, education and urban-rural inequality and maternal healthcare service usage in Malawi. BMJ Glob Health 1(2). https://doi.org/10.1136/bmjgh-2016-000085
Zandi S, Golbaten Mofrad K, Moradi Faraj A, Salehi G (2021) Energy, exergy, exergoeconomic, and exergoenvironmental analyses and multi-objective optimization of a CPC driven solar combined cooling and power cycle with different working fluids. Int J Thermodyn 24(2):151–160. https://doi.org/10.5541/ijot.873456
Zhu X, Wu G (2021) Green product diffusion: the impacts of asymmetric retailers' strategic product decisions. RAIRO—Oper Res 55(3):1459–1486. https://doi.org/10.1051/ro/2021072
Zucaro A, Ripa M, Mellino S, Ascione M, Ulgiati S (2014) Urban resource use and environmental performance indicators. an application of decomposition analysis. Ecol Ind 47:16–25. https://doi.org/10.1016/j.ecolind.2014.04.022

Chapter 2
Overview to Renewable Energy Systems

Abstract Energy is essential for the continuation and development of human existence at every stage. The majority of a nation's energy requirements may be satisfied by commercial sources such as fossil fuels (coal, oil, and natural gas), hydroelectric electricity, and nuclear power. These are treated as the three primary sources for the commercial energy sources. Solar energy from the sun is all around us every day, much like the air we breathe. The sun's life-giving qualities of warmth and light have long been appreciated by us. Solar energy may now be utilized for specific purposes inside homes and businesses thanks to advancements in technology. The pace at which energy is demanded is steadily and alarmingly increasing year after year. With solar technology, we can collect energy from the sun for immediate use. Heat is one method in which solar energy may be stored, and the light of the sun can be harnessed to provide electricity for a business or a residence. Because solar radiation varies across the country, system performance assumptions should be dependent on location and orientation. This work provides a brief overview of factors to be considered for the study energy resources.

Keywords *Energy* · *Energy efficiency* · *Renewable* · *Energy intensity*

2.1 Overview

Energy that is obtained from natural sources and that can be replenished at a faster pace than it can be used is referred to as renewable energy. For instance, the sun's rays and the wind are two examples of such sources that are continually being refilled. Renewable sources of energy are abundant and may be found all around us. The majority of the earth's areas normally have access to various types of renewable energy sources. In comparison to other forms of renewable energy sources, solar radiation has an even wider range of potential applications. During certain times of the year, the capacity factors of some types of renewable energy, such as geothermal and ocean thermal energy, are significantly higher. Solar, wind, hydro, biomass, geothermal, ocean, and ocean thermal are the several types of renewable energy systems that may be differentiated from one another. There are several different energy conversion technologies that can turn these renewable sources into goods

D. Kumar, *Renewable Energy Scenarios in Future Indian Smart Cities*, Advances in Geographical and Environmental Sciences, https://doi.org/10.1007/978-981-19-8456-3_2

that may be put to good use. Both thermal and electrical work may be done using the energy that is derived from the sun. Conventional solar thermal systems are ideally suited for use in residential settings, where they may supply not only energy but also heating, cooling, and even potable water. Solar thermal systems often involve concentrated solar power, which may power a variety of industrial operations that call for temperatures between medium and high. In addition, concentrated photovoltaic systems are capable of producing far better conversion efficiencies from sunlight to power. The surplus heat has the potential to be put to use in a variety of different processes, including cooling, desalination, and others. In a smart city, sophisticated ways are needed to manage and coordinate supply and conversion technologies and demand applications. The proliferation of sensors and integrated computational intelligence algorithms may aid with energy systems integration. Suitable approaches are required to manage complicated networks of players, frequently with opposing purposes, while deciding design and operational choices for systems over a broad range of characteristics and timelines (Dwivedi et al. 2021). This assessment focuses on methodologies in the major application areas together with relevant implemented examples, while emphasizing some of the significant issues being encountered and identifying potential routes for the industry. A thorough description of a city's project framework illustrates design phases and control structures. The report summarizes the present status of computational intelligence in smart energy management and suggests solutions to overcome hurdles (Chen 2011; Kumar 2019; Mujtaba et al. 2020). Innovative technology and integrated solutions help cities design and execute sustainable energy systems, according to a new study. We design sustainable energy systems as part of smart city policies, define goal scenarios and master plans for energy system transformation, and develop and support implementation projects (Kim 2018; Mohamed et al. 2021; Nayan et al. 2013). Cities and towns innovate future-proof lifestyles. On the cusp of the digital era, they allow people and corporations to build smart goods, services, and solutions to enhance their quality of life. Smart cities prioritize sustainable energy. The metropolises are responding to air pollution, traffic congestion, a lack of parking places, obsolete infrastructure, severe weather events related to climate change, increasing energy costs, demographic shift, and greater rivalry for firms, employment, and well-trained personnel (Chettry and Surawar 2021; Shah et al. 2021).

Renewable energy deployment on a large scale has become a priority issue for governments and policymakers in all industrialized economies, emerging markets, and other developing countries over the last decade. Moving from talk to action, on the other hand, is frequently far more difficult than anticipated. These changes have ramifications and necessitate a shift in policymaking, administration, and economic management at all levels. For such transformations, a thorough understanding of the various capacities is required to provide a smooth transition to an energy system (Grochulska-Salak et al. 2021; Leal Filho et al. 2019; Thirumurthy et al. 2012). This is what has sparked an interest in working on solar energy research. The methodology offered here was designed to provide a solid evidence base at the local, regional, and national levels for establishing an enabling environment for solar energy adoption in any given place. Renewable energy is a cornerstone of sustainable practices and economies, providing opportunities to meet the needs of the poor and accelerate long-term economic growth, particularly in developing countries (Kasaeian et al. 2020; Leal Filho et al. 2019). To meet the needs of developing countries, the global energy sector must expand dramatically. Renewable energy will be critical in enhancing millions of people's access to modern energy services, reducing economic risk by guaranteeing energy security, and helping to mitigate climate change risk by lowering emissions (Joffe and Smith 2016). The photovoltaic (PV) effect converts sunlight into energy, which is used to generate solar electricity. Many solar cells are used to make solar panels. A solar electric system or a solar array is created when solar panels are threaded together in series and integrated with additional components. Solar cells power all satellites in space, making them accountable for global communications. They also have an impact on everything from calculators to remote highway signs, as well as homes, businesses, and big power plants (Ciocănea et al. 2019; Nfaoui and El-Hami 2018).

2.2 Energy Crisis in India ~ (Current Status and Trends)

Although India's energy consumption is rising, the country's per capita consumption of 3629 kWh is still very low when compared to the average of 17,620 kWh for the whole world. The average energy consumption per person in India is one-sixth of what it is in the rest of the world, one-half of what it is in China, and much less than one-twentieth of what it is in the USA (Kaur and Garg 2019; Supe et al. 2020). There are around 580 million individuals in India who do not have access to electrical power. Even though theoretically 90% of the population is within reach of the electricity network, only 43% of the population is connected since the cost of connection is too expensive for many people who are living in poverty (Cronin-de-Chavez et al. 2019; Harish et al. 2020). Even homes with ordinary earnings are often discouraged from connecting to the grid owing to the low quality of service, which includes frequent blackouts and outages. Typically, this is the case in the USA. The nation is ranked sixth on the list of the world's greatest energy consumers and is a net importer of energy. India is a massive nation that is home to a diverse collection of natural assets (Dixit 2019; Sawle et al. 2016). The consumption of energy is caused by a variety of fuels, each of which has a distinctive pattern of application. The installed generating capacity in India increased from a pitiful 1362 MW in 1947 to 101,630 MW in early 2001, representing an approximately 75-fold growth from that initial low point. The amount of power that is required in India is enormous and is growing quickly. For India to effectively manage the economic and environmental difficulties presented by expanding energy demand, the country needs to get an understanding of how these and other variables influence energy consumption in the different sectors of the economy (Ataei et al. 2015; Meisen et al. 2006; Sharma and Harinarayana 2017). To evaluate the potential for improvements in energy efficiency or to monitor the execution of rules that are currently in place, it is necessary to collect data that is both precise and consistent (Dheeraja et al. 2022; Kaur and Garg 2019).

As the cost of electricity continues to climb and as worries about the environment intensify, consumers interest in solar energy systems is growing. A solar energy system may reduce monthly energy costs, ensure that the lights continue to work in the event of a power outage, and contribute to the expansion of our economy by creating new employment and solar-related companies (Alberto et al. 2006; Ataei et al. 2015). The Jawaharlal Nehru National Solar Mission is a significant initiative that is being undertaken by the governments of India and its states. The goal of this mission is to promote development that is less harmful to the environment while simultaneously addressing the issue of energy insecurity in India. As a consequence of the efforts put forward by both the state and the federal governments, the researcher was able to identify a variety of programmes that aim to maximize the use of solar cells to produce electrical energy (Marques-Perez et al. 2020; Rn et al. 2014; Sharma and Harinarayana 2017). In every sector of the economy, the energy demand is increasing at a rate that is outpacing the rate at which it can be supplied. This demand–supply imbalance may be considered the source of the growing energy problem.

Solar energy is among the most accessible and easily available natural energy resources, with significant potential to reduce greenhouse gas emissions from a variety of sources. These are used for a variety of applications, including heating, refrigeration, and lighting, and are powered by electricity provided by PV systems and steam power plants. As a result of industrial development and population growth, there has been a massive increase in worldwide energy demand in recent years. Since the early 2000s, a bottleneck has existed due to increased demand for energy, particularly from liquid fuels, and constraints on the rate of fuel supply. This has resulted in an energy crisis (Ali Shah et al. 2021; Luna-Rubio et al. 2012). One of the major bottlenecks (or price increases) in the availability of energy resources to an economy is an energy crisis. This paper outlined a power generation, consumption, and demand scenario, as well as projected power requirements (Wang et al. 2021; Xiao and Wang 2019). It also tried to figure out how energy was stressed and how it was linked to the environment, poverty, and long-term viability. The energy problem has manifested itself in several ways in the context of India's economic development like as:

(a) *Oil Shortage:*

Reserves of mineral oil in India are only 0.3% of the world's known oil reserves and rising oil prices had led to rising general prices in India, consequently affecting the transport sector.

(b) *Coal shortage:*

Coal reserves are quantitatively and qualitatively poor. Coal shortages adversely affect the generation of electricity, therefore, restricting economic development.

(c) *Power shortage:*

A severe shortage exists in the generation and distribution of electrical power in India. Power shortage affects industrial and agricultural production seriously.

2.3 Significance of Research

2.3.1 *Scientific Relevance*

Developing quasi-renewable energy (such as solar energy) will fund worldwide efforts to address the issues faced by climate change, which is caused by massive increases in greenhouse gases, particularly CO_2. Alternative energy sources have been adopted by public interest groups in an attempt to reduce CO_2 emissions. For a variety of reasons, research into smart solar grids is required to enable a new, sustainable, and flexible energy system. We may now create and sell electricity to the grid from rooftop solar panels, potentially transforming us into "prosumers." Geospatial technology has reconsidered the system of solar renewable energy system assessment and deployment, highlighting the significance of aspects such as functionality, sustainability, and usability. While little research has begun, it has received little attention. Solar energy systems can now be customized to meet specific needs. With no greenhouse gases and unlimited free energy, we can save ecosystems and livelihoods.

2.3.2 *Societal Relevance*

To support any economic and commercial operations, an uninterrupted energy supply is required for vital sustainable growth. Sustainable development, in its true sense, should be widely disseminated throughout all three aspects—social, economic, and environmental. The most crucial feature of all of these categories is energy. Not only do production and consumption patterns on a local and global scale determine societal activities, but they also influence key environmental challenges such as pollution, the greenhouse effect, and desertification. Solar resource potential provides the technological, economic, and organizational factors needed to perform a non-essential role. The final judgement on solar power generation's long-term viability is also backed up by social acceptance. In addition, India is rapidly progressing towards obtaining the coveted position of a developed country.

2.4 Research Problem

Solar Energy Modelling and Smart Grid Feasibility Assessment: To research, analyse, and construct solar radiation models that consider sun fractions, topography, and cloud cover, among other factors. Data and strategies for integrating renewable energy assessment and planning approaches are lacking. Solar incoming energy and its incidence in capture units vary seasonally and annually, depending on their locational characteristics. Modelling and assessing the investment needs for solar photovoltaic systems necessitate a methodology. In addition, the study is required to address the synergistic combinations of energy potentials, as well as their spatial allocation details for the natural sound environment.

2.5 Research Objectives

The current study uses geospatial technology along with other analysis approaches as a tool for solar energy modelling, with a focus on photovoltaic energy potential, to analyse the viability of smart solar networks. The objectives of the study include:

1. Modelling solar energy, data processing, and solar map (Sol-Map) generation.
2. Quantifying solar insolation for photovoltaic (PV) yields estimations.
3. Spatio-Temporal and Hyper-Temporal (*S*-*T* and *V*-*T*) variability analysis of solar insolation.
4. Feasibility analysis of smart solar grid (SSG) for power generation.

2.6 Research Questions

The study is meant to respond to the following research questions:

1. What are the approaches to model solar energy at a given geographic location?
2. What are the various approaches of solar radiation data processing and map product generation?
3. How to quantify the amount of solar energy potential for identification of potential energy harvest sites?
4. What are the spatial–temporal and hyper-temporal differences in the amount of solar insolation throughout the local seasons?
5. How much electricity can be generated with the suggested energy systems on an hourly, daily, weekly, monthly, and annual basis?
6. What is the demand–supply scenario of the proposed solar grid project and how much surplus energy can be propagated to electricity supply grids?

2.7 Outline of the Experimental Area

The experiment was carried out over the whole of the Kalaburagi (formerly Gulbarga) district, which is situated in the northern part of the state of Karnataka. The region has a total area of 10,951 km^2 and is situated on the Deccan Plateau. Its coordinates place it between 76°.04′ and 77°.42′ E Longitude and 17°.12′ and 17°.46′ N Latitude (see Fig. 1.1). The height above sea level might range anywhere from 300 to 750 m in most cases. The district is crossed by two major rivers, Krishna and Bhima. Kumar (2017) says that black dirt is the most common variety in the district. In terms of population growth and the degree of urbanization, the district is rapidly changing. There are three distinct seasons in this area. Summer lasts from late February through mid-June, with temperatures ranging from 38 to 44 °C. The southwest monsoon follows, lasting from late June to late September and bringing high rains of up to 750 mm (27 to 37 °C). It is subsequently followed by dry winter weather (11 to 26 °C) through mid-January (Fig. 2.1).

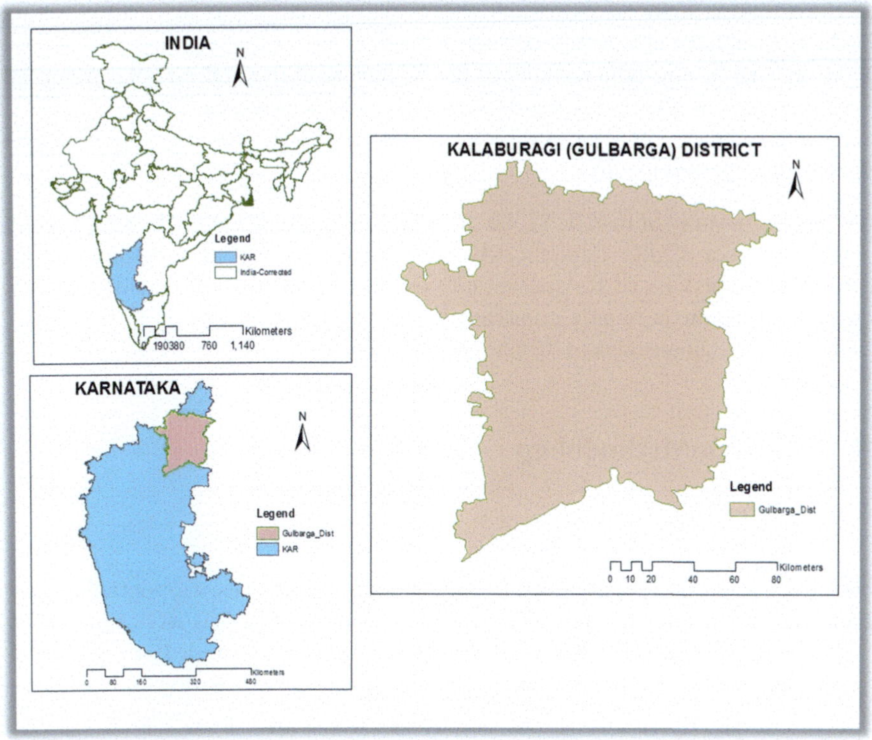

Fig. 2.1 Location of the experimental study area

Why the Kalaburagi (Erstwhile Gulbarga) as a research location?

To broaden the scope of the study being conducted in the region, a particular site cum location in an experimental district is selected. The following factors made the Kalaburagi district a good choice for the location:

1. The absence of fuzzy borders is a self-contained zone.
2. A distinctive and highly varied nature of the environment, with several biological and landscape zones to explore.
3. The goal is to become an eco-island, with a significant interest in sustainable living, self-sufficiency, and the use of renewable energy.
4. A diverse selection of proposed and/or approved planning applications for each of the renewable energy scenarios that are currently under consideration.
5. The purpose of this illustrative example is to give insights and answers to issues about the viability of installing renewable energy sources for the district, which are about current recommendations to test the simulation.
6. The author has a strong familiarity with the region as well.
7. The potential of the resource; yet, what about its applicability? This gives rise to the possibility of several interesting follow-up questions in the study, such as,

(a) Even though there is significant potential for solar resources, why does the district not have any permitted solar farms?
(b) Given that South India has one of the regions with the highest solar potential, is there land in the region that is ideal for solar development?
(c) How do chances for solar installation compare with the use of energy resources that are supplied locally?

A feasibility study of installed solar grids is also conducted at the Central University of Karnataka (CUK) Campus, which is situated on 654 acres of land on the north-western outskirts of Kalaburagi city (about 21 km from the main city). The geographical extent is largely constrained between the latitudes of $17°41'$ N and $17°44'$ N and the longitudes of $76°65'$ E and $76°68'$ E.

2.8 Research Methodology

2.8.1 Materials Used

In this investigation, GCPs obtained from handheld GPS devices, a topographic sheet produced by the Survey of India at a scale of 1:50 K, SRTM DEM/BHUVAN DEM, and very high-resolution (VHR) satellite data obtained from Google Earth were used (for site validations). Additional open archive data sets for processing and validation are available from BHUVAN NOEDA (NRSC), Landsat Archive, and Earth Explorer (USGS). The following parameters were used to access data on solar radiation and other meteorological factors from several accessible archives.

1. Global solar irradiance, direct normal incidence (DNI) solar irradiance and diffuse solar irradiance.
2. The rate and direction of the wind.
3. The temperature of the air at the ambient level and the relative humidity.
4. The pressure in the atmosphere and the amount of precipitation.

2.8.2 Technology and Information Sources

The information was gathered from multiple open source or educational websites along with in-situ data. The ancillary facts were accumulated from books, journals, reports, and websites. The various information sources extensively referred to include:

1. Online gateways and databases *(ResearchGate)*.
2. Ministry of New and Renewable Energy.
3. Online Journals *(e.g., Solar Energy, Energy, Journal of Solar Energy Engineering)*.
4. Books in the context of infrastructure management, trends in energy sciences.
5. Reference texts from ESRI, ERDAS Inc. publications and manuals.

2.8.3 Solar Radiation Data

Solar Radiation Data Handbook for India by Anna Maine is the first comprehensive document that provides the data for many stations in India. The well-known sources of solar radiation data include:

1. In situ measurement of solar radiation.
2. Acquired from the empirical models.
3. Computed from physically based models.
4. From the satellite observations *(Meteosat data).*

2.8.4 Software's and Tools Used

1. Energy Plus 8.0
2. Geospatial Toolkit *(NREL)*
3. System Advisor Model *(SAM)*
4. HOMER *(Hybrid Optimization of Multiple Energy Resources)* Software
5. RET Screen Plus Software
6. ESRI ArcGIS 10. x (*for vector data analysis*)
7. ERDAS 201x (*for image processing*)
8. MS Office 201 × for documentation purposes.

2.8.5 Research Methods

Geospatial technology research can be seen as a fascinating combination of theory and observation. Analytical and observational methodologies were used to complete the current study project.

(a) ***Analytical Research Design***
Refers to the estimate and choice of methodologies that are engaged in the process of data analysis. A significant part of the research design is comprised of the analytical design. Both the choice of data to collect and observe, as well as the output in its ultimate form, are impacted by the method of analysis that is selected. The terms mathematical models and statistical algorithms are often used to refer to analytical procedures. Analytical methods are also used to comprehend the link between information that has been visually interpreted or digitally processed and the real state or phenomena that is taking place on the ground to evaluate the situation.

(b) ***Observational Research Design***
The crux of scientific investigation, as well as the method through which new scientific knowledge is produced, is the juxtaposition of empirical data with

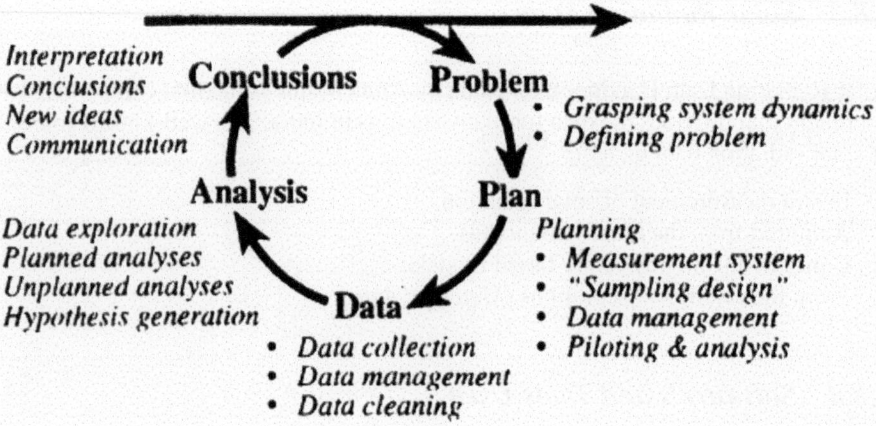

Fig. 2.2 Schematic for the PPDAC methodology

conceptual frameworks derived from theory. The quantitative relationship analysis significantly depended on mathematical and statistical approaches, which made it possible to deduce the existence of causal connections by drawing inductive conclusions from the findings of the mathematical and statistical procedures. In a more theoretical sense, quantitative data and the use of mathematical and statistical analysis are coming to be seen as instruments that may help preserve the objectivity of scientific inquiry (Fig. 2.2).

2.8.6 Research Framework

The entire research was operated on the encryptions of **PPDAC** methodology which are assumed to be very crucial in the field of Geostatistics and Geo computations. The five key elements of **PPDAC** are: Problem, Plan, Data, Analysis, and Conclusion and the complete sequence of activities can be summarized as:

2.8.7 Analysis and Evaluation Methods

Standard statistical packages like SPSS and Microsoft Office Excel were used to examine the statistical relations, perform cross-tabulation, and plot the graph, line, charts, etc. Apart from statistical analysis, several other analyses were being performed including feasibility analysis (technical, economic, and operational) and SWOT analysis.

2.8.8 *Research Setup*

The different stages for accomplishing the current research include:

***Phase 1*: Developing Research Framework**

Defining Research Philosophy, Prototype, and Principles.
Selecting Proper Methodology (like Inductive, deductive or technical approach).

***Phase 2*: Research Design**

Sampling Design (like Ground or Remote Sampling).
Observational Design (i.e., theory and Observation).
Analytical and Operational Design (i.e., variables & relationships, validity & reliability).

***Phase 3*: Data collection**

Data for Geospatial Research (i.e., In situ data, Remote Sensing or other ancillaries data).
Factors Influencing the Selection of Geospatial Data *(i.e., resolution, electromagnetic spectrum).*
Factors Influencing the Selection of Ancillary Data *(i.e., ground truth or other ancillary data).*

***Phase 4*: Pre-Processing and Data Organization**

Image Enhancement, Georeferencing, Filtering, Noise Reduction.
Listing, Categorization, Tabulation, Segmentation, Stratification.

***Phase 5*: Data Analysis and Output Generation**

Data Analysis and Data Mining.
Visual Image Processing.
Digital Image Processing.
Information Output and Level of details.
Converting Geospatial Data into Information.

***Phase 6*: Creating and Validating Solar Energy Radiation Model**

Creating Models using Software and Available Ancillary Data.
Validation of Model Output with Available Data.

***Phase 7*: Feasibility Analysis**

Strength (**S**), Weakness (**W**), Opportunity (**O**), and Threat (**T**)
Technical, Economic and Operational Feasibility

***Phase 8*: Documenting the Research**

Report Generation
Research Paper
Dissertation/Thesis

2.9 Assumptions and Design Controls

1. There were a meagre number of errors in the data used for the research which was comparable to the authentic data.
2. Data derived from different satellites using modelling techniques were assumed to be authentic due to constraints of the financial support to procure instruments and to carry out full-fledged research. The generated data set can be used for forthcoming research also.

The earth's life is fuelled by energy. Energy is required for the survival of all biotic components in this biosphere. Everything, directly or indirectly, is dependent on the energy of the sun. Humans and animals both require the sun's heat and light to live. It is required by plants to carry out the photosynthesis process. Later, when the remains of these animals and plants are fossilized for millions of years, different types of fossil fuels are created based on the type of fossil, the amount of pressure, and the quantity of heat. These energy sources are indeed a tremendous source of energy, but they are finite, and the time it takes to generate them is far longer than the rate at which they are used in daily life, making them a non-renewable source of energy. The combustion of fossil fuels, such as coal, crude oil, petroleum, and natural gas, releases greenhouse gases, causing climate change and environmental destruction. Renewable energy is derived from renewable sources of energy that are naturally replenished in a human timeframe, such as solar energy, hydroenergy, biomass energy, geothermal energy, tidal energy, and wind energy, in addition to non-renewable energy. Because of its low carbon emission characteristics, renewable energy is often known as "clean" energy. Conservation and efficiency of energy are becoming significant goals for which the world must strive. Daily, a significant amount of energy is lost, which should be preserved not just to save money but also to ensure the long-term viability of our natural resources. The ability of humans to transfer energy from one form to another has resulted in the rapid development of our civilization. A certain quantity of energy is squandered during this process. Using efficient machinery to reduce the quantity of energy wasted is a step towards energy efficiency.

Consumption of energy has continued to rise indefinitely. Other than expanding urbanization, unsustainable development, industrialization, transportation, manufacturing, and many other factors, no one can be responsible for this growing trend. With rising global energy consumption, efficient and sustainable energy use is essential; otherwise, humanity will face severe consequences, some of which are visible and alarming, such as climate change, biodiversity loss, abnormally rapid melting of glaciers, ozone layer depletion, and negative effects on public health. Previously, renewable energy was used to power daily tasks such as transportation, lighting, and heating. Sailboats and waves were both propelled by the wind. Windmills used the strength of the wind to grind grains. Biomass was used for a variety of purposes in the past. However, for thousands of years, humans have relied on cheaper and dirtier energy sources such as coal to harness energy. The usage of renewable energy is expanding, from rooftop solar panels used at home to large wind farms, as more innovative and less expensive ways to retain renewable energy like solar and wind energy become available. Entire rural populations rely on these resources only for heating and lighting in certain circumstances. Non-renewable energy sources are significantly more destructive than renewable energy sources in terms of environmental impact. Renewable energy sources such as geothermal, hydropower, and biomass, on the other hand, have substantial environmental consequences. Various elements, such as the technology utilized, the geographical location, and others, influence the type and severity of environmental repercussions. The magnitude and possibility of these environmental consequences of renewable energy use must be understood so that they can be used wisely to minimize or mitigate these consequences.

References

Alberto C, Arau S De, Ame L, Ame L, Das N, Soumendu C, Ansar C, Division ES, Jacoby HD, Supervisor T, Marks DH, Systems E, Reader T, Whittle AJ, Dutta D, Sekhar C, Arnab D, Jain M, Dawa D, Bothwell J (2006) Overview of renewable energy potential of india. Model Earth Syst Environ 2:1–20. https://doi.org/10.1016/j.compenvurbsys.2015.03.002

Ali Shah SA, Longsheng C, Solangi YA, Ahmad M, Ali S (2021) Energy trilemma based prioritization of waste-to-energy technologies: Implications for post-COVID-19 green economic recovery in Pakistan. J Cleaner Produ 284. https://doi.org/10.1016/j.jclepro.2020.124729

Ataei A, Rashidi R, Nedaei M, Kurdestani E (2015) Techno-economic viability of a hybrid wind and solar power system for electrification of a commercial building in Shiraz, Iran. Adv Energy Res 3(4):251–263

Chen YY (2011) New innovation patterns? Lessons learned from digital technology industries. In: PICMET: portland international center for management of engineering and technology, Proceedings. https://www.scopus.com/inward/record.uri?eid=2-s2.0-80053362357&partnerID=40&md5=d5298cfc6ec9bab56f2d95a39a328803

Chettry V, Surawar M (2021) Assessment of urban sprawl characteristics in Indian cities using remote sensing: case studies of Patna, Ranchi, and Srinagar. Environ Dev Sustain. https://doi.org/10.1007/s10668-020-01149-3

Ciocănea A, Dragomirescu A, Tofan B, Toti M (2019) Pedestrian-level ventilation in an urban environment adjacent to a river channel: A case study for Bucharest city—Romania. In: GCNIDASM-CCBF Balan MCGCNIDASMCCBF, Georgescu A (eds) E3S Web of conferences, vol 85. EDP Sciences. https://doi.org/10.1051/e3sconf/20198507007

Cronin-de-Chavez A, Islam S, McEachan RRC (2019) Not a level playing field: a qualitative study exploring structural, community and individual determinants of greenspace use amongst low-income multi-ethnic families. Health Place 56:118–126. https://doi.org/10.1016/j.healthplace.2019.01.018

Dheeraja K, Padma Priya R, Ritika, T (2022) Optimal real-time pricing and sustainable load scheduling model for smart homes using stackelberg game theory. In: Lecture notes on data engineering and communications technologies, vol 99. https://doi.org/10.1007/978-981-16-7182-1_22

Dixit A (2019) Electric vehicle infrastructure market sustainable growth in indian scenario, needs and suggestions. Int J Innovative Technol Exploring Eng 8(11 Special):242–245. https://doi.org/10.35940/ijitee.K1048.09811S19

Dwivedi SK, Roy P, Karda C, Agrawal S, Amin R (2021) Blockchain-based internet of things and industrial iot: a comprehensive survey. Secur Commun Netw. https://doi.org/10.1155/2021/7142048

Grochulska-Salak M, Nowysz A, Tofiluk A (2021) Sustainable urban agriculture as functional hybrid unit—issues of urban resilience. Buildings, 11(10). https://doi.org/10.3390/buildings11100462

Harish S, Singh N, Tongia R (2020) Impact of temperature on electricity demand: evidence from Delhi and Indian States. Energy Policy, 140(111445). https://doi.org/10.1016/j.enpol.2020.111445

Joffe H, Smith N (2016) City dweller aspirations for cities of the future: How do environmental and personal wellbeing feature? Cities 59:102–112. https://doi.org/10.1016/j.cities.2016.06.006

Kasaeian A, Razmjoo A, Shirmohammadi R, Pourfayaz F, Sumper A (2020) Deployment of a stand-alone hybrid renewable energy system in coastal areas as a reliable energy source. Environ Prog Sustain Energy 39(3). https://doi.org/10.1002/ep.13354

Kaur H, Garg P (2019) Urban sustainability assessment tools: a review. J Clean Prod 210:146–158. https://doi.org/10.1016/j.jclepro.2018.11.009

Kim M-S (2018) Research issues and challenges related to Geo-IoT platform. Spat Inf Res 26(1):113–126. https://doi.org/10.1007/s41324-017-0161-z

Kumar D (2017) Surface temperature variability analysis of an urban area using Landsat ETM+ thermal images. Asian Geogr. https://doi.org/10.1080/10225706.2017.1322992

Kumar, D. (2019). Hyper-temporal variability analysis of solar insolation with respect to local seasons. Remote Sens Appl Soc Environ 15. https://doi.org/10.1016/j.rsase.2019.100241

Leal Filho W, Salvia AL, Paço AD, Anholon R, Gonçalves Quelhas OL, Rampasso IS, Ng A, Balogun AL, Kondev B, Brandli LL (2019) A comparative study of approaches towards energy efficiency and renewable energy use at higher education institutions. J Cleaner Prod 237. https://doi.org/10.1016/j.jclepro.2019.117728

Luna-Rubio R, Trejo-Perea M, Vargas-Vázquez D, Ríos-Moreno GJ (2012) Optimal sizing of renewable hybrids energy systems: a review of methodologies. Sol Energy 86(4):1077–1088. https://doi.org/10.1016/j.solener.2011.10.016

Marques-Perez I, Guaita-Pradas I, Gallego A, Segura B (2020) Territorial planning for photovoltaic power plants using an outranking approach and GIS. J Cleaner Prod 257. https://doi.org/10.1016/j.jclepro.2020.120602

Meisen P, Quéneudec E, Yuan M, Nara A, Bothwell J, Ramirez L, Zink R, Dorner W, Stoeglehner G, Sharma P, Harinarayana T, Journal E, Studies E, Vol M, Division ES, Jacoby HD, Supervisor T, Marks DH, Systems E, Ramkrishna B (2006) Overview of renewable energy potential of India. Model Earth Syst Environ 2(October):1–20. https://doi.org/10.1016/j.compenvurbsys.2015.03.002

Mohamed M, Othman A, Abotalib AZ, Majrashi A (2021) Urban heat island effects on megacities in desert environments using spatial network analysis and remote sensing data: a case study from western Saudi Arabia. Remote Sens 13(10). https://doi.org/10.3390/rs13101941

Mujtaba G, Rashid Z, Umer F, Baloch SK, Hussain GA, Haider MU (2020) Implementation of distributed generation with solar plants in a 132 kV grid station at Layyah using ETAP. Int J Photoenergy. https://doi.org/10.1155/2020/6574659

Nayan F, Islam A, Mahmud S (2013) Feasibility study of smart grid in Bangladesh 1531–1535. https://doi.org/10.4236/epe.2013.54B290

Nfaoui M, El-Hami K (2018) Extracting the maximum energy from solar panels. Energy Rep 4:536–545. https://doi.org/10.1016/j.egyr.2018.05.002

Rn HK, Shafiullah GM, Maung A, Oo T, Stojcevski A (2014) Feasibility study of standalone hybrid power system modeled with photovoltaic modules and ethanol generator for victoria. Int J Eng Sci Invention 3(8):54–69

Sawle Y, Gupta SC, Bohre AK (2016) PV-wind hybrid system: a review with case study. Cogent Eng 3(1):1–31. https://doi.org/10.1080/23311916.2016.1189305

Shah, A., Garg, A., & Mishra, V (2021) Quantifying the local cooling effects of urban green spaces: evidence from Bengaluru, India. Landscape Urban Plann 209. https://doi.org/10.1016/j.landurbplan.2021.104043

Sharma P, Harinarayana T (2017) Solar energy generation potential along national highways. Int J Energy Environ Eng 4(16):1–13

Supe H, Avtar R, Singh D, Gupta A, Yunus AP, Dou J, Ravankar AA, Mohan G, Chapagain SK, Sharma V, Singh CK, Tutubalina O, Kharrazi A (2020) Google earth engine for the detection of soiling on photovoltaic solar panels in arid environments. Remote Sens 12(9). https://doi.org/10.3390/RS12091466

Thirumurthy N, Harrington L, Martin D, Thomas L (2012) Opportunities and challenges for solar minigrid development in rural India. Sept

Wang Z, Cai J, Wang Q, Wu SC, Li J (2021) Unsupervised discovery of thin-film photovoltaic materials from unlabeled data. Npj Comput Mater 7(1). https://doi.org/10.1038/s41524-021-00596-4

Xiao F, Wang J-Q (2019) Multistage decision support framework for sites selection of solar power plants with probabilistic linguistic information. J Clean Prod 230:1396–1409. https://doi.org/10.1016/j.jclepro.2019.05.138

Chapter 3
Advancement in Renewable Energy Scenarios

Abstract Looking at what's already been done may inform us which approaches have been tried and tested, which have been successful, and which have been met with resistance in the past. You'll therefore be in a stronger position to choose a technique that provides reliable responses to your research questions. Learning about the existing information in the subject requires that you first review the literature. It's a crucial part of the learning process, playing a role in every stage. It has been determined that the capacity to do labour is the essence of energy. Thermal energy, electrical energy, gravitational energy, luminous energy, chemical energy, nuclear energy, and so on are only a few examples of the many various manifestations of energy. The motion of an object is the driving force behind the distinction between kinetic and potential energy. The first rule of thermodynamics, or the conservation of energy, states that energy cannot be created or destroyed, but that it can be transformed from one form of energy to another. Let's pretend an apple possesses potential energy while it hangs from a tree; this energy would be converted to kinetic energy when the fruit fell. The potential energy created by the arrangement of subatomic particles in the nucleus of an atom is called nuclear energy. In a similar vein, the kinetic energy of water is harnessed and transformed into electricity, which is then used to power appliances and provide thermal and visual comfort. Humans get the kinetic energy they need to do work from the chemical energy they get from the food they eat.

Keywords *Clean energy* · *Energy* · *Efficiency* · *Environmental impacts* · *Management*

3.1 Overview of Earth's Energy Resources

Rapid urbanization raises energy consumption and harms the environment. Urban activities should focus on transitioning to renewable and low-carbon energy sources and enhancing integrated energy solutions. Such programmes should analyse the present energy system, renewable energy, and low-impact alternatives. Integrating renewable and alternative energy sources with the grid is key to designing high-energy performance neighbourhoods (Dwevedi et al. 2018). Renewable energy systems (RES) covers electricity-generating, heat-generating, and energy-efficient

D. Kumar, *Renewable Energy Scenarios in Future Indian Smart Cities*, Advances in Geographical and Environmental Sciences, https://doi.org/10.1007/978-981-19-8456-3_3

technologies (e.g., PV, small wind turbine, fuel cells) (Huber et al. 2014). Waste to energy (WTE), biomass, and cogeneration/tri-generation are alternate energy systems (AES). Methods to analyse potential synergy between RES and AES technologies, and their suitable mix within a specific urban development, as well as the design and characteristics of this development, must be explored (Hossain et al. 2020; Sawle et al. 2016). The research in this article seeks to establish optimal energy resource mixes within an integrated urban energy system for various types and designs of building clusters. This research proposes a technique to optimize the energy resources mix for a grid-tied hypothetical mixed-use neighbourhood, focusing on the technical features of energy resources and their theoretical application to the neighbourhood (Kannan et al. 2021; Kaur and Garg 2019). PV and roof-integrated solar thermal collectors (STC), wind turbines, and waste to energy are explored. Optimization incorporates hourly power, heating, and hot water needs (Abnavi et al. 2019; Islam 2012; Sinha et al. 2017).

Because some of the earth's resources are finite, people are looking for resources that can be replenished in a short amount of time. Inexhaustible renewable sources are the name given to these resources (Azapagic 2004; Cooper et al. 2022). Solar energy, tidal energy, wind energy, geothermal energy, and other limitless renewable sources are examples. These types of resources are currently being used to generate cleaner energy. Governments are also incentivizing people to use these resources by offering discounts on instalments and permits. Solar electricity is one of the most frequent. It may be installed in your home and is simple to use and maintain, and the government makes technology more accessible to you. Solar power is the conversion of solar ray energy into electricity by photovoltaics or focused sunlight, or a combination of both, to achieve the desired effects. Photovoltaic cells are solar cells that can convert sunlight into electricity. Therefore, solar energy is affected by the weather. Solar energy, on the other hand, is one of the purest sources of energy available to mankind. Solar photovoltaics is currently a low-cost, low-carbon technology for harvesting the sun's endless renewable energy (Nixon et al. 2013; Quader et al. 2016; Yang et al. 2016). Because the sun's energy supply is endless, photovoltaics is used to generate solar energy. Many multinational corporations, government infrastructure, schools, colleges, and other institutions have erected solar farms and solar panels to create cleaner energy in recent years (Dovì et al. 2009; Wu et al. 2013). Although it may not be practical in some areas, such as those with heavy humidity and rainfall. According to the International Energy Agency, most solar panel installations are expected to be in China and India.

When it comes to solar panel installations in 2025, the International Energy Agency predicts that China and India will lead the way. Although solar energy has long been known to be one of the most promising renewable energy sources, its potential has been mostly untapped until lately. In all, the earth absorbs just 2% of the sun's energy (in the form of visible light and infrared (heat) radiation), with the other 98% being radiated back into space (Meisen et al. 2006; Noh et al. 2015). The intensity of solar radiation (irradiance) at the top of earth's atmosphere remains almost constant (solar constant) at the average distance that the planet is from the sun, with a measured value of 1366 W/m^2 0.3%. The solar constant is defined as this quantity. Only around half of this energy is thought to reach earth's surface. When we talk about "global solar radiation", we're referring to the sum of all the short-wave radiant energy emitted by the sun's disc and all the energy that's been diffusely diffused by the atmosphere and clouds and made its way over a certain horizontal region in each amount of time. Daily monitoring of global solar radiation is useful for determining the total solar energy at each location. This helps make sure that seasonal and diurnal shifts are considered. The two types of solar radiation that reach earth's surface are known as direct and diffuse. "Global radiation" refers to the sum of all rays, both direct and diffuse, that strike a flat surface.

Direct radiation is the part of the sun's rays that bypasses the atmosphere entirely to reach the surface, whereas diffuse radiation is the part that is scattered by atmospheric components such as molecules, aerosols, and clouds. Direct radiation is responsible for the formation of shadows, while diffuse radiation is responsible for the formation of the skylight. Reading over previous works may be taxing on your time, nerve-wracking, and frustrating, but it also has the potential to be rewarding since it opens your mind to new concepts. The process of reviewing the relevant literature may be time-consuming, intimidating, and irritating; nevertheless, it is also rewarding since it (a) clarifies and focuses the research topic, (b) enhances the methodology and broadens the knowledge base, and (c) contextualizes the results.

Technology examines the relevant published works to see what previous research has discovered. To conduct an effective search for relevant published works in our area of expertise, we need to first have a broad comprehension of both the topic area and the issue that we want to investigate to define appropriate search criteria. After that, we compile a bibliography for this sizable region by using information gleaned from a range of open-source or educational websites, in addition to data collected on-site and supplemental facts (accumulated from books, journals, reports, and websites). Several information sources include (a) online gateways and databases (Scopus, Web of Science, PubMed, IEEE, Research Gate), (b) the ministry of new and renewable energy (website, reports, manuals), and (c) online journals (e.g., solar energy, energy, journal of solar energy engineering), (d) books in the context of infrastructure management, trends in energy sciences, and (d) reference texts from ESRI, ERDAS Inc. publications and manuals (Table 3.1).

Table 3.1 Properties of the sun responsible for PV energy generation

Radiant energy output	3.94×10^{26} W
The radius of the sun	6.960×10^{8} m
Mean sun–earth distance	149.6×10^{6} km
The radius of the earth	6.378×10^{6} m
Energy intercepted by earth	1.8×10^{17} W
Solar constant[a]	1367 Wm^{-2}

[a] *The solar constant is the amount of the incoming solar radiation per unit area, measured to the outer surface of the earth's atmosphere, in a plane perpendicular to the rays*

Producing solar energy involves converting the energy contained in sunlight into a usable form through the photovoltaic (PV) effect. When exposed to sunlight, solar panels convert it into photovoltaic (PV) power via a series of on/off cycles. Sun panels have several individual solar cells working together. A solar electric system, also known as a solar array, is formed when solar panels are linked in series and then coupled with additional parts. Communications on a global scale are handled by solar-powered satellites in space. In addition to affecting homes, companies, and even large power plants, they also affect calculators and remote highway signs. There are two types of solar panels that are commonly used: crystalline silicon and thin film. Solar energy harvesting is based on a basic idea (Gevelt and Holmes 2015; Rn et al. 2014; Zhou et al. 2022). The photovoltaic effect is a physical or chemical process that can turn photons into electricity in tiny cells. Solar cells or photovoltaic cells are the names given to these cells. In this process, photovoltaic cells play a critical role. When a photovoltaic cell is exposed to sunlight, the photovoltaic effect is a chemical or physical process that generates an electric current or voltage. In general, photovoltaic cells are made up of two types of semiconductors: *n*-type conductors and p-type conductors. A *p–n* junction is formed by joining these semiconductors together. Before moving on to the working principle, there are a few terminologies and processes that must be grasped to fully comprehend the operation of photovoltaic cells.

Although crystalline silicon still dominates the market for solar cells, thin-film technologies like amorphous silicon and cadmium telluride are gaining ground. Recently, solar companies started producing solar energy using plastic and aluminium foil (Gevelt and Holmes 2015).

3.1.1 Commonly Used Solar Radiation Measuring Instruments

There are two methods to quantify solar radiation: by ground-based observations at points on the earth, or by measurements made by geostationary satellites. While the latter approach makes use of geostationary satellites, the former entails measuring the energy reflected by the system (earth and atmosphere) in various wavelength bands (Xiangdong et al. 2021). Since the satellite source is the source of the input radiation for the models that were built as part of this research endeavour, the study of the related literature exclusively focuses on satellite-based data (Utrillas et al. 1998). On a horizontal surface, regular measurements of diffuse, global, and direct radiation will be made by an entity like the national meteorological office. A pyranometer will be used to obtain these measurements. This organization is also in charge of overseeing and administering a network that consists of all the meteorological stations in a nation (Mulugetta and Drake 1996; Ramachandra and Subramanian 1997). This section will provide a quick overview of some of the devices that are used most often to measure a variety of radiation components. The least amount of focus will be paid to the instruments that are used to measure the many descriptors and parameters that are used in solar radiation measurements. A detector or sensor is necessary for any apparatus that measures radiation. The calorimetric, thermomechanical, thermoelectric, and photoelectric thermometers are the primary kinds of instruments used to measure temperature (Panidi et al. 2020; Zhang et al. 2021).

Pyrheliometer—When measuring the direct sun radiation in W/m^2, the normal incidence angle is used. It is composed of a collimator, which is a small hollow tube with an optimal aperture of around 6°. With such a precise aperture, the collimator can exclude almost all scatter radiation while yet capturing the whole solar disc in its field of vision. The collimator houses a thermopile detector that has a quick reaction time and several junctions inside of it. A sun-tracking equatorial mount that is electrically driven is often used to support a pyrheliometer while it is being used.

Pyranometer—It is possible to calculate the total quantity of solar radiation (W/m^2) that a complete hemisphere is exposed to (diffuse and direct, combined). In a typical setting, this hemisphere spans the whole sky dome. A hemispherical glass dome shields a flat, blackened thermopile detector positioned on a base. A pyranometer that is slanted at an angle or a "spherical pyranometer" is two more methods that may be used to detect ground-reflected radiation. The detector gives a response based on the overall power regardless of the spectral distribution of the light that is being absorbed. The heat that was created because of the black disc's ability to absorb radiation is transferred from the thermal resistance to the heat sink. The difference in temperature that is measured across the thermal resistance of the disc is then translated into a voltage that a computer can interpret. The dirt that accumulates over time must be removed from the glass dome, thus it is necessary to clean it regularly. The body of the pyranometer contains silica gel crystals, which prevent moisture from penetrating the instrument.

Pyranometer with a shading device—Measurement of diffuse solar radiation (W/m2) is accomplished by blocking the sun's direct rays with an appropriately sized shade ring or disc. The former is referred to as a shadow band, which is a stationary device (to the east–west axis) that is oriented or positioned parallel to the polar axis of the earth. To consider the many shifts in the sun's declination angle, it has to be altered several times. In addition to the two primary kinds of radiation instruments described above, the following sorts of devices are used in the measurement of radiation from the sun.

(a) ***Albedometer***—The value of albedo is determined by measuring both the total solar radiation and the solar energy that is reflected. To assure accurate measurements, this apparatus is often installed at a height of several meters above the ground.
(b) ***Pyrometer***—Long-wave radiation may be measured if the appropriate equipment is used. Only infrared light with wavelengths between 3 and 50 μm is permitted to pass through the dome, which has been mirrored to reflect as much of the sun's short-wave radiation as is physically feasible. In most cases, it is used in combination with other forms of shade to ensure that exposure to direct sunlight is minimized.

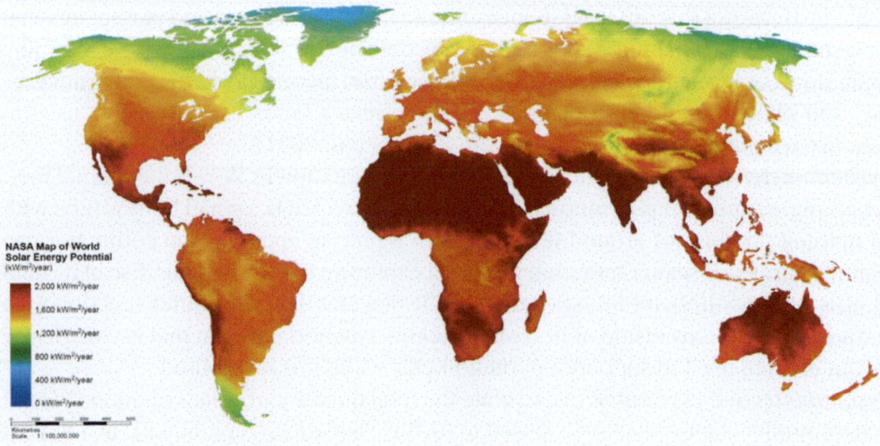

Fig. 3.1 World solar energy potential map

(c) ***UV meter***—As its name suggests, it can identify the portion of the electromagnetic spectrum that is visible from the surface of the planet (UV-A: 0.351–0.400 micro m or UV-B: 0.280–0.315 micro m).

All the instruments are included in the category of broadband instruments. Sky scanners, sun photometers, multi-filter shadow band radiometers, rotating shadow band spectrometers, spectroradiometers, interferometers, and grating spectrometers are all examples of spectrum instruments.

The distribution (also known as the energy budget) of the sun's incoming solar energy. As can be seen, the great majority of the radiation is taken in by the earth's land and water. As a consequence of this, the energy balance is maintained via the different materials by being absorbed, radiated, or reflected in an analogous method. The potential for solar energy generation all around the world is shown in Fig. 3.1. In addition to that, it incorporates a graphic representation of the current solar resource supply on the Indian subcontinent.

3.1.2 Solar Radiation Modelling

Several studies have emphasized the need of modelling solar data. This section looks at some of the most important landmark models in the history of solar radiation (also known as correlations or regressions). The two forms of irradiance are short-wave and long-wave irradiance. The latter is exemplified by infrared irradiance, whereas the former is exemplified by global and, of course, its two components. Short-wave irradiance is also known as broadband or spectrum irradiance, and it is mimicked in this way. Although a short overview of spectrum models will be relevant here, the focus will be on the latter because this study is primarily focused on broadband radiation modelling. The assessment will largely focus on models and modelling methodologies designed for satellite-based observations because all the measurements utilized in this study are satellite-based. As a result, a knowledge overview of satellite-based models should be provided. In the context of this work, modelling solar radiation data comprises three important aspects. To obtain "clean consistent datasets", solar radiation energy modelling requires processing and filtering. The data processing method is handled after the modelling step. The third step is to assess the first-aspect datasets to recommend the "best" model method. Finally, it entails quantifying energy measurements gathered in various data formats, as well as the creation of solar maps as a final step.

The several varieties of solar radiation models that were discovered in the study are shown in Fig. 3.2. The sheer nature of satellite-based models makes them fundamentally separate from one another. Although these models cannot be sold under the same umbrella, they do pose additional obstacles and complexities. When it comes to models that are based on the earth's surface, often known as terrestrial or ground-based models, accuracy typically improves as the model's complexity grows from left to right. Radiation models that are currently suggested may be found among the categories that have been highlighted. It is important to point out that a model will often include features from several different groups. It is important to keep in mind that, unless otherwise specified, when the term "radiation" is used, it will always indicate the radiation that is assessed (or predicted) on a horizontal surface. Before this, several different types of radiation models have been presented. Over time, they have been subjected to a variety of tests, and, as a result, a lot of researchers have improved upon them. The results of a comprehensive classification of these models may be broken down into two distinct categories:

(a) ***Models which estimate diffuse or direct radiation, or both, based on geographic and topographical characteristics or global radiation alone/satellite-based models.***
(b) ***Models that estimate both diffuse and direct radiation simultaneously.***

The term "global into diffuse and direct/beam irradiance decomposition models" is used to refer to models of this kind. There is a large amount of variation in the quantity of solar radiation that reaches the surface of the planet owing to geographical and climatological variances (Shiva Chandra and Reshma 2022; Sinha et al. 2015). Solar radiation monitoring networks that have a density that is proportional to the geographic complexity are difficult to create (Alijani et al. 2020; Guleria et al. 2019). Because of this, solar radiation models that are constructed using information obtained from metric satellites are becoming more and more significant. These satellites have a very fine resolution in space and can survey a large area. Modelling approaches vary from those that are purely subjective to those that make use of some of the finest complex radiative transfer programmes.

Perez et al. developed a model based on "monitoring the dynamic range of satellite image pixels and assigning irradiance values proportional to the relative brightness of the pixels". The model was created using this way. Users who seek to estimate the quantity of solar energy received at the earth's surface using satellite data, on

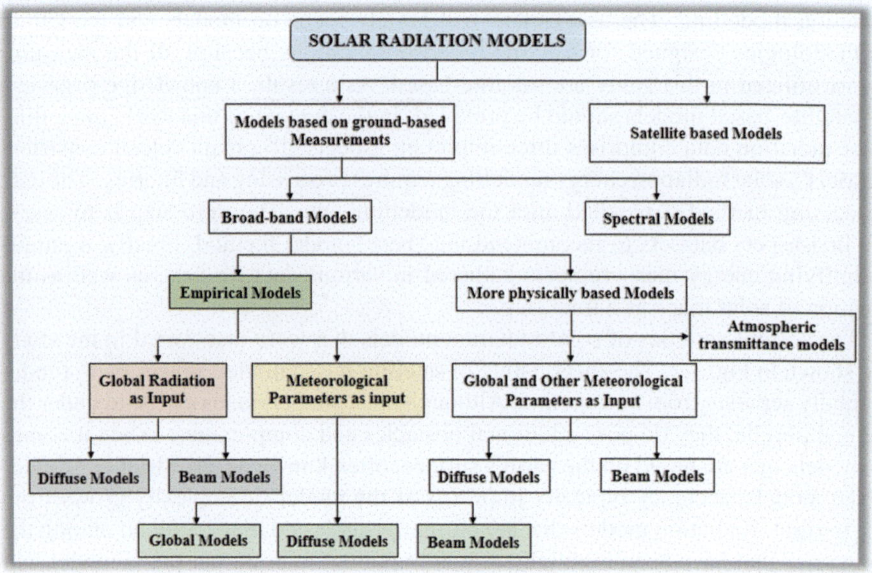

Fig. 3.2 Network diagram presenting hierarchy of solar radiation models

the other hand, encounter several difficulties (Sinha et al. 2015). When compared to ground-based observations, Vignola et al. (2005) discovered that the satellite-based model was unable to distinguish between frost on the ground and low-lying fogs and/or clouds on clear winter days. This was identified when the model was compared to ground-based data taken on clear winter days. They reached the opinion that to carry out in-depth modelling, ground-based observations must be employed to augment satellite data (Kabisch et al. 2019; Kumar 2019a). This was the result that they got. When compared to the data that was collected, it was discovered that their model's anticipated values for diffuse radiation had poor accuracy. In addition to the fact that it is difficult to adjust satellite-based models within the constraints of the earth's current radiation and energy budget, there are other considerations at play (Al-enezi et al. 2011; Xiao et al. 2014).

(c) ***Models that estimate the various components of solar radiation using meteorological data, with or without considering the effects of global radiation.***

These models are simple to apply since they only need one observable input variable, which is global radiation. The fundamental problem with these methods is that they are unable, on their own, to explain or account for the variation in diffuse fractions. The decision to classify it also considers a variety of other considerations.

3.2 Current Scenario of Solar Energy Systems

A solar electric system typically consists of solar panels, an inverter, a battery, a charge controller, wiring, and a support structure. Grid-connected solar power systems, grid-connected solar power systems with battery backup, and off-grid solar power systems are the three most popular kinds of solar power systems (stand-alone). Every one of them has a different set of skills and applications.

(a) ***Grid-Connected***

Solar panels in this setup are connected to the electrical grid maintained by the local utility company. This configuration is intended to supplement the conventional source of electricity provided by utility companies. Grid-connected systems include solar panels, an inverter, a junction box (which connects the wiring from the solar panels to the home's breaker panel), a power meter (which shows how much energy is produced and consumed), and a disconnect switch (which, for safety reasons, prevents the system from sending power to the grid during power outages). Grid-connected systems may also transmit electricity to the grid (this is called islanding).

(b) ***Grid-Connected in conjunction with Battery Backup***

It is quite like a system that is linked to the grid in that it has a "battery bank" to store the power that is collected from the solar panels. It features a "battery bank" to store the energy that is gathered from the solar panels, which makes it relatively comparable to a system that is connected to the grid. The energy that is stored in the batteries may be utilized when there is no power available. Solar energy must be gathered, transported to the breaker box, and then fed into the home's electrical system via the battery bank. Solar panels, an inverter (which converts DC solar electricity to AC), a battery bank (for storing power), a charge controller (which prevents the battery from overcharging), a junction box (which connects the solar panel cabling to the home's breaker panel), and a power meter make up this kind of system. The solar energy must be converted from direct current to alternating current via the inverter (which displays the amount of power used, produced, and stored in the battery bank).

(c) ***Stand-Alone System or Off-Grid System***

Off-grid systems are not connected to the utility grid and are often found in rural locations since the cost of connecting to the utility grid is higher than the cost of purchasing an off-grid system. Since the cost of connecting to the utility grid is more than the cost of acquiring an off-grid system, off-grid systems are more common in rural areas. Backup fuel sources, like generators or modest wind systems, are employed when the solar power stored in the batteries is insufficient to meet the demands of the residence (Negi and Mathew 2014; Thirumurthy et al. 2012). When the power goes out, islanding is avoided thanks to a system consisting of solar panels, an inverter (which converts DC electricity into AC energy), a rectifier (which converts AC electricity into DC energy to get the most out of a system), a charge controller (which prevents the battery from being overcharged), two junction boxes (one for the wiring of the solar panels and the other for the wiring of the generator), and a power meter (which displays the amount of power used, produced, and consumed by the system).

3.3 Photovoltaic System Architecture

Light is made up of photons, which are essentially bundles of radiant energy. When photons reach a PV cell, they have the potential to be reflected or absorbed (transmitted through the cell). The generation of electricity is limited to only those photons that have been taken in. When photons are absorbed by atoms in a solar cell, their energy is then transferred to the electrons that make up those atoms. Silicon, which is the second most common material on earth after sand, is generally used in the construction of solar cells. Silicon is typically used in the construction of solar cells. One piece of silicon has had a trace amount of boron incorporated into it, and, as a result, it has an increased propensity to attract electrons. It is referred to as the *p*-layer because of its generally favourable tendencies (Adelaja et al. 2011; Anand et al. 2017). A trace quantity of phosphorous was put into the other piece of silicon, and as a result, it now possesses an abundance of free electrons. The tendency of the n-layer to release electrons with a negative charge is where the term "*n*-layer" originates from.

3.4 Solar Panel Technologies

There are two traditional forms of solar panels used to generate solar electricity: crystalline silicon and thin film; however, these new goods may take several years to reach consumers.

(a) ***Silicon Solar Panels***

Crystalline panels, which are made up of a single giant crystal, and polycrystalline panels, which are made up of multiple smaller crystals, both of which are solid (made from blocks of silicon that contain many small crystals). Silicon solar panels, although the most efficient currently available, are also the costliest. In addition, they outperform all other types of panels while operating in dim conditions. Silicon solar panels have an average efficiency of roughly 33%, whereas polycrystalline solar cells are somewhat less efficient than single-crystal solar cells. This is the case even though the efficiency of single-crystal solar cells is somewhat higher.

(b) ***Thin-Film Solar Panels***

Large glass or ceramic panels are covered with a tiny layer of silicon and other materials that may be utilized to produce solar energy (about the thickness of a human hair). Silicon solar panels have a higher initial cost but generate energy at a higher efficiency level than thin-film solar panels. Due to the solar panels' relatively thin design, they are well-suited for use as solar shingles and tiles, which are integrated into various types of buildings. Some of the most promising thin-film materials include amorphous silicon, cadmium telluride, and copper indium selenide. The degree of efficiency varies from ten per cent up to nineteen per cent.

c) ***Future Solar Panels***

Panels made of lightweight foil, collectors made of plastic, and hybrid solar electric/solar water heating systems are some examples of new solar materials. Modern hybrid systems heat water using the hot air that is collected from solar panels. Except for hybrid systems, the vast majority of newly developed materials are not now accessible to families. The International Energy Agency (IEA) recognizes four primary applications of photovoltaic electricity that are higher than or equal to 40 W.

1. ***Off-grid solar energy systems or domestic solar energy systems***—delivering electrical electricity to residences and communities that are not linked to the grid.
2. ***Off-grid solar energy system or non-domestic solar energy systems***—to provide electricity to industrial applications that are located away from the grid, including water pumping, alarm lighting, and environmental data collecting.
3. ***Grid-connected distributed PV energy systems***—to provide electricity to a structure or other load linked to the grid.
4. ***Grid-connected centralized energy systems***—developed to deliver a replacement to conventional methods of centralized energy production or to enhance energy utility distribution systems.

The synoptic overview of solar panel technologies is shown in Table 3.2, along with some short information.

Table 3.2 List of solar panel technologies

Sl. No	Technology type	Area (m^2/kW_p)	Efficiency (%)	Cost of panel
1	High-performance hybrid silicon (HIT)	6–7	17–18	\$1.40–2.00 per watt,
2	Monocrystalline silicon	7–9	12–15	\$0.45–1.50 per watt,
3	Polycrystalline silicon	7–10	11–14	\$0.35–1.35 per watt,
4	Thin-film CIS	9–11	9–11	\$0.35–1.20 per watt,

Source Various internet sources

3.5 Prerequisites for Solar Energy Assessment

3.5.1 *Space and Orientation*

For every kilowatt of electricity that is generated by a solar electric energy system, about 100 square feet of shade-free, south-facing roof or yard area is required. There may be a space requirement of 175 square feet per kW for thin-film devices. It is not a good idea to install solar power systems in locations that are exposed to the east, west, or north. The angle at which your roof is pitched is also vital for collecting the sun's rays; however, the precise requirements for doing so change depending on where you live.

3.5.2 *Energy Modelling Algorithm*

Irradiance readings must be taken at predetermined intervals to properly evaluate the amount of energy that solar power system (SPS) plants produce. Information on solar radiation may be obtained from several different sources. There are two different kinds of measurements: satellite measurements and ground-based measurements. Satellite-based irradiance data sources include organizations such as NASA, Meteonorm, NREL, 3Tier, ISHRAE, and Weather Analytics. Other providers include Solemi. The following are some of the solar energy modelling algorithms that have seen the most widespread application:

(a) **Bird Clear-Sky Model** shows the solar radiation of a clear sky as seen through the eyes of a bird. It discusses the fundamental methods and factors that cause solar radiation to be absorbed and scattered in the atmosphere. It demonstrates how solar radiation is affected by the height of the sun and the state of the atmosphere. It calculates the direct and diffuse solar radiation components at ground level (Myers 2005a, b).

(b) **Global Horizontal Irradiance Clear-Sky Models** compute the terrestrial solar radiation under a clear sky as a function of the solar elevation angle, site height, aerosol concentration, water vapour, and any other environmental factors that are relevant (Reno et al. 2012).

3.5.3 *Solar Geometry*

It is possible to indicate the position of the sun from two different points of view. The first angle is called the solar azimuth, and it is defined as the angle that points counter clockwise between the sun and the cardinal direction of true north. It is determined according to the horizontal projection of the position of the sun onto the surface of the earth. The combination of the two angles provides vital information on the path taken by sunlight as it approaches a particular item or structure. Keeping this information in mind, solar collectors and other equipment should be positioned within 20° of the sun on either side or perpendicular to it. Including a mechanism that can adjust to the angle at which the sun is shining may allow us to better regulate the angle at which the incident light strikes the surface of the collector. For instance, light shelves that are hinged may have their angle altered to get the desired effect.

The tilt of the Earth

The axis that runs from pole to pole on earth has tilted at an angle of about 23.5° for the horizontal plane that its orbit occupies. A difference of 47° in the peak solar height angle is responsible for the hemisphere-specific differences that occur between summer and winter. When seen from the northern hemisphere, the sun is always on the south-facing (equator-facing) side of the building, and the angle at which it strikes the house varies during the day. Because of the inclination of the earth, the height of the sun in the sky is at its lowest in the northern hemisphere during the winter, while it is getting close to its highest point in the summer. This is a significant consideration that should be addressed before the installation of windows and shading in a building. During the winter months, when the sun is at its lowest point in the sky, beneficial radiation and natural light can more easily enter the inner region.

Slope and Aspect

When doing terrain analysis, the slope, aspect, and surface curvature of a cell are all determined by utilizing the elevation data of the cells that are immediately next to it. Authors such as Skidmore and Jones, as well as Zhou and Liu, have compared several methods that may be used to determine slope and aspect (Zhang et al. 2016). Because the resolution is a function of the slope, it is usually a good idea to describe the spatial resolution that was utilized to calculate the slope and aspect of the terrain. There will be a point where tangents (slopes) that are perpendicular to the east–west and north–south directions will pass through the elevation. These two tangents provide two components known as $\partial z/\partial x$ and $\partial z/\partial y$. These components may then be used to determine the general direction of the slope as well as the aspect of the slope. The gradient is a vector variable with components that are the partial derivatives of the surface in the x and y dimensions. Using methods that distinguish between components running east–west and north–south, the overall slope and aspect of a 3 × 3 grid may be determined with the use of the following formulas.

$$\tan S = \sqrt{\left(\frac{\partial z}{\partial x}\right)^2 + \left(\frac{\partial z}{\partial y}\right)^2} \tag{3.1}$$

$$\tan A = \left[\frac{\left(\frac{-\partial z}{\partial y}\right)}{\left(\frac{\partial z}{\partial x}\right)}\right] \tag{3.2}$$

Zhou and Liu describe another algorithm for calculating aspects, as follows:

$$A = 270^\circ + \arctan\left[\frac{\left(\frac{\partial z}{\partial x}\right)}{\left(\frac{\partial z}{\partial y}\right)}\right] - 90^\circ\left[\frac{\left(\frac{\partial z}{\partial y}\right)}{\left|\frac{\partial z}{\partial y}\right|}\right] \tag{3.3}$$

3.5.4 Parameters of Solar Panels for Different Applications

(a) **Insolation on horizontal surface (kWh/m²/day)**

The quantity of solar radiation and other forms of electromagnetic energy that make contact with the surface of the planet. Solar radiation may also be referred to as global solar radiation or total solar radiation. There is information supplied on the average, as well as the minimum and maximum percentage differences.

(b) **Diffuse radiation on horizontal surface (kWh/m²/day)**

Amount of electromagnetic energy (solar radiation) that strikes the surface of the earth under conditions of clear sky and indirect radiation from the sun's beam being blocked at the surface of the earth by a shadow band or tracking disc.

(c) **Direct normal radiation (kWh/m²/day)**

Amount of electromagnetic energy (solar radiation) at the surface of the earth when the radiation from the sky is blocked and the surface is flat and perpendicular to the beam of light from the sun.

(d) **Insolation at 3–hourly intervals (kW/m²)**

The quantity of solar radiation and other forms of electromagnetic energy that make contact with the surface of the planet.

(e) **Insolation clearness index (dimensionless)**

The amount of sunlight that makes it down to the surface of the earth from the uppermost layers of the atmosphere.

(f) **Insolation normalized clearness index (dimensionless)**

It is possible to calculate the insolation clearness index without taking into account the zenith angle.

(g) **Clear-sky insolation(kWh/m²/day)**

The average quantity of sunlight received on days with few clouds (cloud cover of less than 10%).

(h) **Clear-sky insolation clearness index (dimensionless)**

The average amount of sunlight received on days with low cloud cover (less than 10%).

(i) **Clear-sky insolation normalized clearness index (dimensionless)**

The clear-sky insolation clearness index has an expression that is not dependent on the zenith angle.

3.5.5 Parameters for Tilted Solar Panels

(a) **Radiation on equator-pointed tilted surfaces (kWh/m²/day)**

To calculate radiation on inclined surfaces, the monthly average insolation measured on a level surface is utilized. The tilt angles are as follows: 0° of latitude minus 15, 15° of latitude, 15° of latitude plus 15, 90° of latitude, and the ideal tilt angle. The ideal angle is what supplies the highest radiation that occurs every month on average.

(b) **Minimum radiation for equator-pointed tilted surface (kWh/m²/day)**

To calculate radiation on inclined surfaces, the monthly minimum insolation measured on a level surface is utilized. The tilt angles are as follows: 0° of latitude minus 15, 15° of latitude, 15° of latitude plus 15, 90° of latitude, and the ideal tilt angle. The ideal angle is what supplies the highest radiation that occurs every month on average.

(c) **Maximum radiation for equator-pointed tilted surfaces (kWh/m²/day)**

When calculating radiation on inclined surfaces, the monthly maximum insolation that occurs on a horizontal surface is utilized as the reference point. The tilt angles are 0° latitude minus 15, 15° latitude, 15° latitude plus 15, 90° latitude, and the ideal angle. The ideal angle is what supplies the highest radiation that occurs on a monthly average.

3.6 Background of Solar Resources in India

3.6.1 The Indian Subcontinent

South Asia is home to the vast subcontinent that forms the country of India. The Arabian Sea, the Bay of Bengal, and the Indian Ocean are the three bodies of water that encompass the peninsula. The Arabian Sea can be found to the southwest, while the Bay of Bengal and the Indian Ocean can be found to the southeast and south, respectively. India spans a range of latitudes from 8.4° north to 37.6° north, and it spans a range of longitudes from 68.7° east to 97.3° east. It occupies a little more than one-third of the space that the USA does thanks to its overall land area of 3,287,590 km^2, which is slightly larger than that of the USA. The distance from the northernmost point to the southernmost point of India is 3214 km, whereas the distance from the easternmost point to the westernmost point is 2933 km. It has a land border that is 15,200 km long and a coastline that is 7517 km long. Combined, these two measures the whole length of its boundary.

The whole nation may be broken up into four separate natural zones. There is a desert area in the northwest, the Himalayas run along the northern boundary, the Deccan Plateau extends through the centre and south of the nation, and the Himalayas run along the northern boundary. The Ganges lowlands are rich and heavily inhabited. Due to the majority of the country's landmass being situated in tropical zones, the weather remains warm throughout the whole year. The Himalayas act as a barrier, protecting people and animals from the icy winds that blow from the north. The influence of the monsoon may be seen in the climate, which is hot and dry for eight months of the year but rainy from June to September of each year. Solar radiation levels in the majority of India are acceptable, ranging from 4 to 7 $kWh/m^2/day$, which is sufficient to meet growing energy needs and cover deficit regions. Has the ability to drastically reduce the amount of kerosene and gasoline that is used for illumination and the production of electricity.

3.6.2 Developments in the Renewable Energy System

In the last decade, India has taken significant action to take advantage of the several advantages presented by renewable energy. Energy resource systems centred on renewable resources are actively encouraged in India, with the medium-term goal of promoting decentralized energy technology powered by renewable resources. Ministry of Non-Conventional and Renewable Energy (MNRE) officials have said that one of their objectives is to boost capacity by 12 GW by 2012. More than 100,000 mW of renewable energy capacity are expected to exist in India (MW). To close the gap between the country's energy demand and supply, this potential must be used methodically and strategically. Presently, renewable energy accounts for around 3.5% of the total installed electric energy system capacity, or about 3700 mW (MW). It was expected that by 2012, this share would rise to 10% of total power generation capacity. India is home to the world's second-largest biogas and improved stove programme as well as the world's fifth-largest wind energy programme.

3.6.3 Solar PV Progress in India

When it comes to sunshine, India ranks high. Solar energy has a huge potential for application in the generating of electricity and the heating of buildings. Many parts of India have over 300 sunny days annually, making it a prime location for the expansion of solar power. India has a wide range of solar energy incidence, with daily averages between 4 and 7 kWh/m2 and annual averages between 2300 and 3200 h of sunshine. There is a huge market for solar power in India. The solar energy market in India has enormous growth potential in commercial and industrial applications. If the conversion efficiency of PV modules is 10%, the country gets enough sunlight to produce more than 500,000 TWh of power annually. India's key markets for photovoltaics (PV), which have a total installed capacity of 107 MW, are telecommunications power, railway networks, the oil and gas industry, the military services sector, and international trade. One of the most essential requirements for comprehensive rural development in terms of infrastructure is the electrification of rural communities. About eighty thousand American villages lack access to the national energy grid. Unfortunately, even with an expansion of the current infrastructure, only 18,000 of these communities will have access to reliable energy. The Tenth National Plan has set a goal of bringing power to 5,000 of these communities.

3.7 A Critical Analysis of Recent Advances in Solar Energy Research

Recently, for a variety of reasons, there has been a lot of interest in the use of models and tools to produce or aid in the assessment, analysis, and optimization of renewable energy and energy efficiency technologies. This interest can be attributed to the models and tools' ability to generate or facilitate the process. The implementation of several technological developments can be done on a global, regional, local, or even project-by-project basis. Each of the models and tools has a variety of layouts that are either designed for consumers or energy experts. Numerous national and international investigations have been conducted in this regard. However, it has been noted that the following studies on this subject are noteworthy:

3.7.1 A Review of the Literature Regarding the Modelling of Solar Energy, Data Processing, and the Generation of Solar Maps

In the context of the renewable energy system in general and solar "photovoltaic" technology in specific, a study of system modelling and simulation was conducted. Additionally, a study of the ideal system modified by static DC-DC converters was conducted (Alami et al. 2014). An analysis of the effect that temperature and irradiance have on the ideal parameters (Power of MPP, VMPP,) of a solar system was carried out to ensure that a photovoltaic (PV) generator could operate at its full potential. MATLAB–Simulink was the programme that was used for both modelling and simulation purposes (solar panel, DC-DC converter command and load).

This research was carried out to map, using GIS, the solar potential of western India. The research examined data on solar radiation and temperature gathered over ten years from twelve different places all around the globe. On the other hand, it was pointed out that employing a greater number of stations and taking into consideration climatographic components (such as sunshine hours, fog, altitude, and so on) that need a direct or indirect impact on solar energy should result in improved conclusions (Sarif and Gupta 2021). The findings, on the other hand, were in line with what was expected from the large solar energy-producing facilities in Gujarat. According to the findings of the research, there is the possibility to construct solar farms in the western part of Gujarat, which receives an average of 6700 MJ/m^2 of solar energy annually (Ajayi et al. 2016).

The transition from scheduled to variable production based on renewable energy sources, from centralized to decentralized generation, and from expensive to cost-free energy carriers are three key developments in the energy sector. Infrastructure planning, energy generation, and storage are also important developments. Because of these developments, the energy infrastructure will need to undergo considerable revisions. Even though the vast majority of these shifts are being driven, in a natural way, by the difficulties and issues presented by geography. The incorporation of geographic information systems (GIS) into energy system models was in its infancy at the time. In this regard, (Resch et al. 2014) examined the shortcomings of previous approaches to using GIS for projects related to renewable energy. They extracted distinct challenges from these previous efforts and finally defined a set of primary future research avenues for GIS-based energy infrastructure planning with a focus on renewable energy (Cannizzaro et al. 2021; Zahid et al. 2020). Future research could focus on the availability of base data and their "geospatial awareness", the development of a generic and unified data model, the use of volunteered geographic information (VGI) and crowdsourced data in analysis processes, and the integration of 3D building models and 3D data analysis to add network topologies to GIS. (Adelaja et al. 2011) presented the insight of incorporating the SolMap project as a component of the SRRA to visualize solar resource yields. It went into detail about various types of risks, including technical risks such as component performance and proper performance, operational risks such as realistic modelling of actual plant operations, simulation model uncertainty, input data processing uncertainty, quality checks, product and report generation without and with gap-filled continuous time-series processes, and many more. In addition to that, it included the evaluation of India's solar resources, the formation of TMY, and the creation of solar maps for the country (Kumar 2019a; Pandey et al. 2012).

The research was done on the topic of modelling urban solar energy with a high spatial and temporal resolution by utilizing a three-dimensional model, a technique was created to evaluate the hourly solar photovoltaic energy production potential that may be found on rooftops in an urban environment (Amani et al. 2020; Kandekar et al. 2021). In the area of the case study that was conducted at Ryerson University in Toronto, an application named Ecotect Analysis 2011 was used to estimate the incident solar radiation that was received by each building's rooftop. In addition, by making use of geographic information systems, orthophotos of the area covered by the case study were scanned to exclude unnecessary rooftop features from the model (Amani et al. 2020; Hall and Zidar 1993; Kara and Keçeli 2017). In the end, a software programme known as HOMER was used to produce hourly solar PV power estimates by using the values that were acquired by the other two software programmes as input parameters. It was discovered that the hourly solar PV production followed a pattern known as a binomial curve, and it was observed that the peak solar producing times aligned with the peak energy usage hours throughout the summer (Dheeraja et al. 2022; Tijani et al. 2014).

In the case of a clear sky, clear-sky models compute the amount of terrestrial solar radiation as a function of the solar elevation angle, the site height, the concentration of aerosols, the amount of water vapour, and other atmospheric factors. Validating a model in clear-sky conditions requires making a comparison of the results of the model to the irradiance that was measured during those periods. Within the framework of this discussion, (Reno et al. 2012) developed an innovative approach for recognizing periods of clear sky within a time series of GHI observations. The performance of selected clear-sky models was tested using data gathered from thirty different sites, for a total of around three hundred site years' worth of data. The researchers looked at how mistakes changed throughout the course of time and place. In terms of the mean error across all locations and times, more complex models were somewhat more accurate than other models in terms of accounting for all of the atmospheric properties. However, simpler models are capable of achieving equal accuracy, particularly at low altitudes.

According to the findings of research that was conducted by (Hassan et al. 2011) on solar energy modelling over a residential neighbourhood in Calgary, Alberta, Canada, solar energy is an abundant source of renewable and sustainable energy with significant potential for lowering emissions of greenhouse gases. This research was conducted by Hassan, Rahman, Haque, and Ali. In this paper, a modelling framework is presented that allows for the prediction of solar energy throughout a portion of the Sandstone residential complex located in northwest Calgary, Canada. The actual daily incoming solar energy was calculated as a function of latitude, year, and the number of daylight hours that were available. High-resolution remote sensing pictures were also used to evaluate the effective roof area for the installation of photovoltaic cells. Strong connections between ground-based observations and estimated real incoming solar radiation were found to exist in Alberta at three different test sites, with r2 values ranging from 0.91 to 0.98. Over the Sandstone section, we concluded that there is a suitable amount of roof surface area for the prospective installation of solar cells (Dixit 2019; Okwanya et al. 2021). According to the findings of the research, it was possible to meet a sizeable portion (i.e., somewhere in the range of 67–100 per cent) of the electrical energy requirements of the residents of the Sandstone village during April and September by utilizing solar cells with an efficiency of 15%. (Ahmad and

Tiwari 2010) reviewed and reported on solar radiation models that described data on solar radiation and the components of solar radiation at a particular place. In the course of the study, each of the global solar radiation models that were available in the scholarly literature was investigated. Several formulae connect total global radiation to other climatic variables such as daylight hours, relative humidity, and the highest possible temperature. (Andreas et al. 2010) published research that demonstrated solar radiation models that were incorporated in geographical information systems (GIS) and acted on 2.5D raster data. On the other hand, models that perform point cloud-based solar radiation modelling are not the best choice for vast study areas. This is because the enormous quantity of point cloud data cannot be handled in the main memory of the computer all at once. As a result of this body of work, a cloud-based and point-based solar radiation model was developed, and it was fully included in the SAGA GIS (Ren et al. 2021; Xu et al. 2021).

(Fargas et al. 2021; Sharma and Harinarayana 2017) specified the basics and principles of mathematical models along with the procedures involved in modelling along with different modelling aspects including predictive vs probabilistic model descriptions. Research utilizing GIS for renewable energy mapping in Maharashtra, India, has been published (Kulkarni and Banerjee 2010). The geographical information system (GIS) was used to identify potential locations for the collection of renewable sources of energy (Cannizzaro et al. 2021; Gambhir et al. 2010). Quantum GIS, often known as QGIS, is an open-source tool that was used to investigate the unpredictability of renewable energy from a geographical point of view. Installed capacity maps were created (Marques-Perez et al. 2020a; Sinha et al. 2017). Data on solar insolation and wind speed were plotted and then interpolated for the whole state using data from a few places. A macro-survey was conducted to determine the state's renewable energy potential (Cannizzaro et al. 2021; Marques-Perez et al. 2020b). For microanalysis, spatial interpolation was used to determine the system's utility.

An operational article on solar energy modelling with ArcGIS was prepared by (Ostapowicz 2009) and assessed the potential of photovoltaic electricity production in two mountain areas in Austria and Poland. In this report, they tried to introduce solar radiation model creation and suitability assessment for solar energy. Consequently, a lab manual was prepared for the GIS-based solar radiation modelling dealing use of GIS for solar radiation modelling (Hochwimmer 2009). The technical description of the research focuses on the availability of data on solar radiation resources to allow distributed grid-tied PV (Kazem et al. 2014; Molina and Espejo 2014; Wafa and Taha 2014). It also discusses the types of data and products that are required to support future analysis and operations of significant grid-tied PV penetration (Myers 2005a, b). The majority of the material that was utilized in this study was obtained from conversations with other researchers who have written in the same study region, as well as analysts, utility representatives, and solar resource data providers (Gils et al. 2017; Kaur et al. 2020).

Measurements and modelling of terrestrial broadband and spectrum solar radiation are essential components for solar renewable energy system evaluation and installation. It was claimed that to calibrate wide band solar radiometric equipment and improve the accuracy of broad band solar radiation measurement (Jiang et al. 2018; Myers 2005a, b). To make it easier to do outdoor pyranometer calibrations, both an improved reference for the diffuse sky and software for calibrating and characterizing radiometers were made available to users. It has been briefly discussed how to compute direct beam, total hemispheric, and diffuse sky radiation using many different broadband solar radiation model methodologies (Salisu et al. 2019; Tanskanen et al. 2021; Yang et al. 2017). The National Renewable Energy Laboratory is responsible for the development of a number of these methods (NREL). Estimating direct beam radiation from global data is done with the use of a programme called the Direct Insolation Simulation Code (DISC). The Meteorological and Statistical (METSTAT) and Climatological Solar Radiation (CSR) models are used to measure solar radiation using meteorological data.

According to (Hofierka and Šúri 2002; Schwingshackl et al. 2013), the most significant factor that influences the dispersion of radiation on both a regional and a local scale is terrain, more specifically relief. Variations in elevation, surface inclination (slope), and direction (aspect), as well as shadows cast by terrain features, all contribute to the formation of strong local gradients. The spatial and temporal variation of incoming solar energy drove the dynamics of a variety of landscape processes, which had a direct influence on human culture. These processes included air and soil temperature and moisture, snow melting, photosynthesis, and evapotranspiration (Anderson et al. 2008; Le et al. 2018; Nandy et al. 2021).

It is possible for empirical connections to account for the regional impact of topography, as indicated by (Adelaja et al. 2011; Dixit 2019). This helps to explain not just why there are two kinds of spatial insolation models, but also why one is point-specific and the other is area-based. The amount of insolation that a location receives is determined by utilizing point-specific models that are derived from the geometry of the surface orientation and the visible sky.

The solar analyst is useful for models that are point-specific as well as those that are area-based. It can generate an upward-looking hemispherical viewshed, which would provide the functional equivalent of a hemispherical (fisheye) picture for each location on a DEM, as described by (Adelaja et al. 2011; Tal and Hager 2015). An accurate insolation map was produced as a consequence of employing hemispheric viewsheds in the calculation of the amount of sunlight reaching each location. The solar analyst can determine the integrated insolation for each period. It can take into account the latitude and elevation of the location, in addition to the surface orientation, shadows cast by the surrounding landscape, daily and seasonal changes in the sun's angle, and air attenuation. It was developed specifically for use with ArcView as a GIS extension (Baniya et al. 2018; El-Zeiny and Effat 2017).

Design issues such as GIS implementation and interface, computational issues, and error propagation have been addressed (Dubayah and Rich 1996). The introduction of strong analytic tools that are part of various GIS settings has made it easier to construct such models. Several difficulties, including the selection of appropriate physical models, modelling languages, data formats, and model interfaces, were discussed in this context. The impacts of terrain and plant canopies were discussed, as well as various methods for getting the data required to drive certain solar radiation formulations (Ashrafi Goudarzi et al. 2019).

By introducing the idea of Satellite-to-Irradiance Modelling—A New Version of the SUNY Model, the researchers at SUNY came up with a novel concept (Ezzine et al. 2014; Nolasco et al. 2008). Within the scope of this article, the most recent iteration of the SUNY satellite model that will be used in an operational setting was presented and verified. The most recent version includes several enhancements, including more effective treatment of the model's dynamic range, improved handling of clear sky, the incorporation of now-casting numerical weather predictions, and an improvement in the model's ability to more accurately depict extreme (clear and overcast) conditions. This newly revised version considerably enhanced performance over a broad spectrum of environmental factors (Luna and Medina, 2021; Yariyan et al. 2020; Ahmad and Tiwari 2010).

3.7.2 Literature Review for Quantifying Solar Energy Yield Estimation

The Ministry of New and Renewable Energy has set up 115 SRRA stations throughout the nation. These SRRA stations are automated solar and meteorological monitoring stations. These stations are located in diverse areas. These stations are recognized by their acronym. Accurate measurements are necessary for the planning and development of solar power installations as well as the performance evaluation of these systems. This is because solar radiation varies throughout the day (Ashrafi Goudarzi et al. 2019; Zhao and Fong 2017). It is common practice to utilize satellite-based irradiation estimates for the continuous geographic coverage of large regions. On the other hand, ground-based observations provide the data of the highest possible quality and are also utilized for verifying, benchmarking, and improving satellite-derived data. As part of the Jawaharlal Nehru National Solar Mission (JNNSM), the Ministry of New and Renewable Energy (MNRE) has accepted the enormous challenge of installing 20 gigawatts (GW) of solar power across the country by the year 2022. This is being done under the IPCC protocol, which aims to reduce the detrimental effects that environmental pollutants have on climate change (Dey et al. 2005; Dovì et al. 2009; Hasan et al. 2021).

(Adelaja et al. 2011; Luqman et al. 2015) performed research to estimate the solar energy potential of residential roofs in the Lahore district as an alternative source of power. A few different roof sections were scanned to establish the total amount of accessible rooftop space for the installation of photovoltaic panels. Calculations of the monthly and yearly solar energy potential for the year 2014 were carried out with the use of GIS models. The energy usage of the civilization was nine times the amount of energy that was planned. If there is an energy deficit in the region, more solar PV power would be used to make up the difference (Bawah et al. 2013; Kaur et al. 2020; Olatomiwa et al. 2015; Rauf et al. 2020).

To provide more accurate estimates of PV potential at the roof level in urban areas, some researchers conducted a study on remote sensing and GIS in planning the photovoltaic potential of urban areas. This allowed them to suitably account for distributed factors using active (LIDAR) and passive satellite/airborne remote sensing techniques and advanced GIS modelling tools (Sadat et al. 2020; Santos et al. 2014). It has been shown that renewable energy resources may be analysed using remote sensing and GIS (Adelaja et al. 2011; Cannizzaro et al. 2021; Moutinho and Madaleno 2021). Globally, it is believed that the geometry of earth's rotation and revolution around the sun causes latitudinal variations in radiation, but regionally and locally, topography (relief) is the primary factor changing the distribution of radiation. High local gradients were caused by a combination of factors, including variation in elevation, surface inclination (slope), and direction (aspect), as well as shadows cast by topographical features.

(Bojanowski 2013; Huang et al. 2021; Luna and Medina 2021) provided evidence that meteorological and satellite data may be used to estimate the amount of solar energy reaching the surface of the planet. The primary objective of the project was to develop and validate methodologies for improving estimates of the daily solar radiation received by Europe's surface. The monitoring agricultural resources unit (MARS) crop yield forecasting system of the European commission's joint study centre was the primary application of this research (MCYFS).

(Adelaja et al. 2011; Basir et al. 2013) analysed satellite remote sensing to identify solar potential locations in Pakistan to develop renewable energy sources and technologies to provide solutions to long-standing energy challenges. By conducting research on cloud cover in Pakistan during June, July, and August (the monsoon season with the most clouds) from 2005 to 2009 and highlighting the use of remote sensing in the process of harnessing renewable energy resources, the objective was to find locations in Pakistan that received the most solar radiation. According to the statistics, Multan and Dera Gazi Khan had the lowest average cloud cover both seasonally and annually. This was the case for both cities. Some researcher has created a toolkit for the organization's GIS, development, environment, and energy teams that included both external and internal applications. Before having a consultation, it was intended to assist prospective customers who had a limited understanding of planning and development by providing online tools that might identify planning boundaries and determine whether or not a site was suitable. While the desktop model was what powered the web tool, it also served as a complementary desk survey tool, which helped in efficiently and effectively responding to site inquiries; as a reference tool before the site visits and proposal stages; when preparing environmental impact or sustainability assessments; and at the strategic planning level, which identified future locations for renewable energy development opportunities (MacKerron and Mourato 2013; Zhao et al. 2020).

The web-based solar cadastre and solar irradiance on building roof estimates that were developed by (Agugiaro et al. 2005; Yi et al. 2022) were designed to estimate solar irradiance on building roofs in complex landscapes. The tops of the buildings were modelled using highly sophisticated automated photo matching methods, which allowed for the creation of very high-resolution geometric models. Models were connected with raster and vector data sources so that they could anticipate the amount of incoming solar radiation that would influence the roofs. Throughout the process, several aspects, such as the impact of shadows thrown by nearby buildings, chimneys, dormers, vegetation, and terrain topography, are taken into account. The effects of the atmosphere, as well as the site's latitude and elevation, as well as the slope, aspect, and slope angle, are all taken into account. The development of a study area, which was later put to use in the process of developing an open-source software solution, took place in a mountainous terrain that had 1250 different types of buildings, including residential, commercial, and industrial structures. The procedure was shown accurately by utilizing data obtained from a pyranometer, and the findings were made available via a prototype of a web-based geographic information system.

This study examines the estimation of hourly solar radiation on the level and sloped surfaces in the western Himalayas (Adelaja et al. 2011; Zhao and Fong 2017). These assessments are essential for evaluating the design of solar energy systems and conducting performance evaluations for a variety of sites, especially those that are located in remote areas. To offer solar radiation data on an hourly basis, a model was constructed to estimate worldwide solar radiation by making use of temperature data and the number of hours of sunlight. This model was developed to give solar radiation data on a global scale. These two estimated values of the hourly solar radiation are compared to the actual values that are measured for each hour to determine how accurate the models are (Ahmad and Goparaju 2018; Reno et al. 2012; Xu et al. 2021). We have also estimated the total solar radiation that falls on surfaces that are inclined and vertical in a variety of directions. It turns out that the estimated values and the measured values are quite close to each other. According to the study's authors, the method presented can be used to improve the accuracy of estimating hourly global diffuse solar radiation for horizontal surfaces and total solar radiation for inclined and vertical surfaces at different orientations. This can be done just about anywhere (Ahmad and Goparaju 2018; Kazem et al. 2014; Tanskanen et al. 2021).

(Ashrafi Goudarzi et al. 2019; Vinod Kumar and Krishnaveni 2020) used the solar energy from existing structures (SEES) and solar analyst (SA) models for calculating incoming solar insolation to determine the surfaces for two separate locations. In a comparison of solar models, the SEES model outperformed the SA model in estimating sun irradiation in an urban context. This was mostly due to the SEES model's computation utilizing observable meteorological data. Final evaluations revealed that adequate roof areas are accessible to install PV systems in both study sites and that their overall potential energy production of them might contribute to energy savings (Gambhir et al. 2010; Ruiz-Arias et al. 2009). Furthermore, the study found that the SEES model, as well as GIS-based solar models in general, was quite useful in estimating solar irradiation in metropolitan areas (Fath et al. 2015; Mohajerani et al. 2017).

Renewable energy on brownfield lands in the state of Michigan was studied as a case study (Adelaja et al. 2010; Iqbal and Iqbal 2019). This research went out to learn more about how renewable energy may be used in brownfields. These brownfields in Michigan have wind and solar resource maps placed on them. These overlays were based on these projected brownfield land capacities. Michigan's brownfield sites have an estimated energy potential of 4320 MW of wind plate capacity and 1535 MW of solar plate capacity, enough to provide 43% of the state's residential power usage (using a 30 per cent capacity factor). More than $15 billion in investment and 17,500 jobs, both temporary and permanent, were expected (Kamau-Devers et al. 2021; Kumar 2019b; Santos et al. 2014). This study analysed solar resources by using GIS and remote sensing techniques. It wasn't completely novel for remote sensing data to predict the quantity of energy that may be created by the sun (Datta et al. 2010).

To estimate daily solar radiation, four state-of-the-art solar radiation models were compared and contrasted). The purpose of this was to find the most reliable model (solar analyst, r.sun, SRAD, and Solei-32). A digital elevation model (DEM) was analysed and compared in a difficult topographic area of the Mediterranean. Radiometric data were obtained over 40 days from 14 sites, and the models' predictions were compared to that data (Ruiz-Arias et al. 2009). Additionally, two alternative DEM spatial resolutions (20 and 100 m) were used to assess their impact. The research discovered that under cloudless skies, the daily solar radiation variability could be computed with an error of less than 10% in the mean bias and an error of about 15% in the root mean square. This was the case even though the error values were relatively high. On the other hand, the findings demonstrated that when foggy conditions are present, the accuracy of some models of solar radiation is greatly reduced. When it comes to the impact of DEM spatial resolution, the data suggest that using a larger spatial resolution improves the accuracy of estimates for challenging topographical locations when the sky is clear. This is the case regardless of whether or not clouds are present (Ji et al. 2020; Kim and Kwan 2021).

(Adelaja et al. 2011; Santos et al. 2014) performed a study on the potential of solar power production in EU member states and candidate countries. During the years 2001–2005, a database on European solar radiation was compiled by using a solar radiation model and meteorological data that were incorporated into a photovoltaic geographic information system (PVGIS). The database, which had a resolution of 1 km and spanned the years 1981–1990, included monthly and yearly averages of global irradiation as well as other climate elements that were connected to it (Adelaja et al. 2011; Hofierka et al. 2014). The database was used to analyse regional and national differences in solar energy resources, as well as to assess photovoltaic (PV) potential in the 25 member states of the European Union and the five candidate countries. This study focuses on three aspects: (1) the expected average annual electricity generation of a "standard" 1 kWp grid-connected PV system; (2) the theoretical potential of PV electricity generation; and (3) the required installed capacity for each country to supply 1% of national electricity consumption from PV. According to the paper, photovoltaics (PV) are already capable of making a significant contribution to a diversified renewable energy portfolio in both the present and the future European Union.

The Canary Islands were the first place in the world to have the computation of PV potential maps introduced by (Monedero et al. 2007). The major input factors that were used to evaluate the PV potential of the Canary Islands were solar radiation data, ambient temperature data from ground stations, and a digital elevation map of the islands. Although there were some modifications made to the models, the procedure that was used was derived from a technique that has been used in the past by several researchers to determine the PV potential in Europe and Africa. When determining the PV potential, the ambient temperature is one of the factors that is taken into account.

The study has been conducted on the assessment of solar energy's potential using GIS to predict the availability of resources spatially and over time. Utilizing geographic information systems (GIS) and remote sensing, a mapping of the resources and demand was carried out across many geographical and temporal scales (RS). This study looked at the possibility of using a geographic information system (GIS) to create a map of the solar potential in the Indian state of Karnataka (Ramachandra 2007). This study provides a picture of the potential by identifying locations of the globe that are suitable for capturing solar energy. These regions were determined based on global solar radiation data. According to the findings of the research, coastal regions of Karnataka that get a larger amount of global solar radiation are ideal for the collection of solar energy. The hypothetical analysis suggests that the regions of Uttara Kannada and Dakshina Kannada get the most amount of radiation from the sun globally (Ramachandra et al. 2017; Ramachandra and Kumar 2009).

The need for spatially continuous mapping of solar radiation was mandated by the introduction of the spatial estimation of global solar radiation by the use of geostatistics (Adelaja et al. 2011; Rehman and Ghori 2000). The geostatistical method was used in this article to estimate the amount of solar radiation that is received in Saudi Arabia. This method included the following five steps: (i) data collection; (ii) univariate analysis; (iii) experimental variogram calculations and model fitting; (iv) kriging estimate; and (v) contour map plotting. To measure variograms, different variogram models have to be fitted to each month of the year. For a grid resolution of 55 × 33 km, estimates were produced at 1500 grid points with a longitude range of 36.58°E to 50.00°E and a latitude range of 17.17°N to 31.33°N. These coordinates correspond to a longitude of 36.58°E to 50.00°E and a latitude range of 17.17°N to 31.33°N, respectively. These values were used in the process of developing solar radiation contour maps for every month of the year. The effectiveness of the method was evaluated for use in calculations at the 41 different known sites by systematically excluding one of these locations from the available data (Baniasad Askari et al. 2014; Rn et al. 2014). According to the results of the error analysis, the highest mean difference between the measured and estimated values was 0.0037 in January, while the lowest was 0.0013 in March and October. The researchers observed that the mean percentage of mistakes ranged from a low of 0.5% up to a high of 1.7%.

3.7.3 *Research Review for Spatio–Temporal and Hyper–Temporal Variability Analysis*

The term "renewable energy resources" refers to those that have a cycle duration that is less than one hundred years, are renewed by nature, and have a supply that is more than the demand for them. Systems that use renewable energy are less harmful to the environment since they depend on natural resources that are always being renewed. It is vital to conduct an analysis of resource availability from both a geographical and a temporal perspective to make the most of the potential offered by renewable energy sources. Both geographic information systems (GIS) and remote sensing (RS) contribute to the mapping of resources and demand on both geographical and temporal dimensions. A geographical database that details the availability and demand of resources would be helpful to the regional energy planning effort.

(Gambhir et al. 2010; Perez et al. 2016) review and analyse recent studies to comprehend, define, and simulate solar resource variability on a spatial and temporal scale. According to the findings of this research, gaining an understanding of the variability of solar energy requires not only defining the temporal and spatial context in which variability is evaluated but also describing a predictable, quantifiable space–time continuum that can smooth out the variability and ranges in length from a single point to thousands of kilometres and in time from seconds to days. Both the effects of solar penetration on the electrical grid and the methods that might be taken to reduce variability were investigated.

Using hyper-temporal remote sensing for scaling between spectral indices and flux tower measurements, (Joyi et al. 2015) measure the spatial and temporal fluctuations of environmental factors as a key problem in earth's system research. This is one of the challenges researchers face when studying the earth's system. The purpose of this study is to evaluate the various methods by which data for a typical temperate forest in southern England may be collected with the necessary geographical and temporal sampling. By examining regional differences, researchers were able to estimate the potential outputs of a study aimed at determining the optimal size of a sample study of photovoltaics. In particular, the impact of these technologies on electric networks was investigated (Rome and Shaddick 2014). This research presents an innovative statistical approach to assessing how sample size affects the accuracy with which PV installations in South Wales, UK, can be shown. It was found that several different characteristics, including size and direction, had a significant impact on PV output profiles; hence, probability distributions were employed extensively across the research region to highlight this.

Recovery, qualification, standardization, and accessibility of the best data from the existing solar resource were the goals of the study daily, monthly, and yearly averages of solar radiation and insolation were the focus of this study's mapping analysis. There was one map that showed the locations of the recording stations, as well as one that showed the yearly and monthly contours. The state of Minas Gerais solar radiation is between 4.5 and 6.5 kWh/m^2, according to the map of daily global sun irradiation and the yearly average calculated project.

Many researchers have created and evaluated the techniques to improve land cover mapping and monitoring for hyper-temporal remote sensing and interpretation. They have researched hyper-temporal remote sensing for land cover mapping and monitoring. The specific aims were: (i) to create a simple technique for characterizing long-duration cloud contamination with a hyper-temporal imagery analysis; (ii) to categorize and map land cover gradient through analysis of hyper-temporal imagery; (iii) to assess a spatio-temporally unambiguous and gradient-based landscape heterogeneity mapping approach for natural and semi-natural landscapes; and (iv) to develop a land cover profile assessment method to abstract hyper-temporal imagery over time.

Interdecadal variability in surface solar radiation over Northwest China and its potential cause was one of the topics that researchers investigated (Lian-tong 2011). Through the use of station measurements spanning the years 1961 to 2003, the study investigated the interdecadal shifts in the seasonal mean surface solar radiation that was measured across Northwest China. These results showed that a rise in low cloud cover over Northwest China may be associated with a reduction in spring and summer surface solar energy. The lack of cloud cover over Northwest China and the circumglobal teleconnection in the northern hemisphere were shown to be associated with abnormally high levels of rainfall in the spring, summer, and winter.

In research on short-term irradiance variability, (Perez et al. 2011) studied the link between irradiance variability at two neighbouring sites as a function of their closeness to one another and the considered variable timescale. When there is a strong connection between two solar systems, the cumulative relative variations of both systems go up, but when there is a weak correlation between them, the cumulative relative variations go down. The research revealed that the connections between distance, timescale, and other variables could be forecasted and were often unaffected by factors such as geographic location or the amount of available sunlight. In addition, the findings demonstrate that the distance between stations that must be reached before the correlation between them can no longer be maintained is a quasi-linear function of the timescale that is being taken into consideration. According to studies conducted in the USA on the regional and temporal variations of the solar resource, the solar resource is constant throughout both time and place (Wilcox and Gueymard 2010). The mapped data presented areas with high and low variability, giving readers a fast visual indication to assist them to pick where to perform measurements and for how long to do so for a certain application.

(Bois et al. 2008) presented research on solar radiation spatial and temporal variations in the Bordeaux winegrowing region that made use of data collected by remotely sensed satellites (1986–2005). The Heliosat-2 technique of collecting data on solar radiation was applied, and for that purpose, the HelioClim-1 database was put to use. This database was developed using pictures that were obtained by the Meteosat satellite. To generate horizontal solar radiation maps with a resolution of 500 m, daily data were interpolated using a method known as conventional kriging. After that, high-resolution daily solar radiation maps with terrain integration were generated using a digital elevation model for the years 2001 to 2005, with a resolution of 50 m. These maps covered the period from 2001 to 2005. A 20-year study of solar radiation in Bordeaux grapes at a low spatial resolution (500 m) indicated a diminishing west-to-east gradient throughout the research. Based on mean daily irradiance values during August and September, the winegrowing districts of Bordeaux were separated into the following three zones: low, medium, and high solar radiation. This initial zoning was scaled up to a resolution of 50 m by utilizing a local correction ratio that was derived from research that looked at solar radiation on inclined surfaces during 2001–2005. The eastern half of the vineyards see large changes in solar radiation, which is discovered to be predominantly controlled by topographical slopes and orientations in the Bordeaux winegrowing region. The fundamental significance and consequence of the research are that solar radiation significantly affects vineyard water balance, grapevine growth, and berry ripening. On the other hand, records from meteorological stations seldom include information on irradiation. In this study, the significance of high-resolution mapping is emphasized. Using an open-source GRASS GIS system, (Hofierka and Cebecauer 2008) carried out a geographically distributed assessment of solar energy resources for several applications in Slovakia. The findings showed that the southern regions of the Slovak territory have the largest potential for solar energy production, with an average daily total of real-sky solar irradiation on a horizontal plane of roughly 3.0 kWh/m^2/day (range from 2.7 to 3.3 kWh/m^2/day over 95% of the area). Mountainous regions in the north have a mixed potential that is significantly impacted by terrain (areas affected by shadows). The amount of solar energy that was accessible showed a significant seasonal variation, peaking in the summer. Horizontal planes only get 15% of the sun's irradiation available in June in December. Solar panel tilt may be changed to increase solar energy availability by up to 30% from June to winter.

Some researchers used data from the National Solar Radiation Database (NSRD) (1961–1990) from four Louisiana stations and data from the Louisiana Agriclimatic Information System (2001–2006) from twenty-five stations to analyse temporal and spatial trends in surface global horizontal solar radiation in the state (LAIS). Similar to worldwide patterns observed by earlier research, a linear decline in surface solar radiation was seen in three of the four NSRD sites throughout the thirty years of data (Perez et al. 2009). One station was the only one with a little upward trend. The relationship between temperature and surface solar radiation was positive for the greatest temperature and negative for the lowest. Despite having a more southerly sun angle, Shreveport has a higher summer solar radiation transmissivity than the other three southern Louisiana sites. Transmission levels in southeast Louisiana (represented by New Orleans) were found to be lower than in southwestern Louisiana as a result of the greater impact of large water bodies in the southeast (represented by Lake Charles). All NSRD locations showed a midsummer decline in transmissivity for a "normal" averaged year. Using information provided by the NSRD, the values from each LAIS station were verified. While the majority of LAIS stations had inadequate data, certain stations were suitable for use in a future study, at least in some regions of the six-year time series (Andalibi et al. 2021; Perez et al. 2009).

A model for topographic variation in solar radiation has been provided inside a GIS framework. The surface gradient (slope), direction (aspect), height, elevation, and location of nearby surfaces all affect the clear-sky shortwave solar radiation. Hilly topography demands a complicated network of stations, whereas generally, the flat ground makes radiation measurement straightforward (Anwar and Deshmukh 2018; Tausendschön and Radl 2021). The model used here uses a digital elevation model to calculate possible direct solar radiation and spread-out radiation, but it may be changed to include cloud cover and the quantity of water that might fall as precipitation in the atmosphere. The model's ability to computationally effectively integrate radiation across broad timescales allowed for the modelling of radiation (Kumar et al. 1997; Laur 2014).

3.7.4 Literature Review for Feasibility Analysis of Smart Solar Grids

One of the best characteristics of the current PV environment is the ongoing decrease in costs. Price reductions are being aided by ongoing improvements in the efficiency of cells and modules, but the main element driving costs down is the sheer volume of goods being produced in cutting-edge facilities with high levels of automation (Wang et al. 2021). In this section, we tried to introduce the idea of the "learning curve" to demonstrate how costs often tend to go down as total production goes up over a broad variety of manufactured goods. because economics often determines how quickly a new technology enters the market. Buyers of grid-connected PV energy systems have understood the global market, and they are interested in learning how much it costs to create solar energy. Consider how the price of a unit of electricity (1 kWh) compares to the price paid by the local utility and if building a rooftop PV energy system appears like a wise investment. Since grid electricity isn't always an option, there are specific requirements for stand-alone PV systems (Abnavi et al. 2019; Shi et al. 2015). Decisions are impacted by environmental factors like noise and pollution as well as comparisons with diesel generators.

> *A group of interconnected loads and distributed energy resources (DER) with clearly defined electrical boundaries that acts as a single controllable entity to the grid [and can] connect and disconnect from the grid to enable it to operate in both grid-connected or island mode.*

(Verma et al. 2015) describe an innovative business model used to provide Uttarakhand, India, with high-quality power from a non-conventional energy source. Early research and simulations including a variety of composite power sources, including solar PV along with battery bank, solar PV with battery bank and generator, and solar PV with DG, were done using the HOMER programme. Under five different community models, the techno-commercial viability of these three hybrids was analysed in terms of load-sharing proposals with solar photovoltaics and batteries, distributed generation (DG), energy production through photovoltaics (PV), annual load consumption, monitoring of excess and unmet energy demand, battery sizing to meet the load during nights, and DG operation when solar energy is no longer available. The financial sustainability of three integrated solar PV energy sources was examined for each community model in terms of annual O&M costs, Levelized cost of energy, and cost summary (Kumar 2017). According to the analysis, a solar PV power plant with a battery bank is the most cost-effective choice for a hybrid system. It is feasible because of cheap operational and upkeep expenses, falling prices for batteries and solar PV modules, and battery pricing at the time of replacement (Gambhir et al. 2010; Li et al. 2014).

According to the power quality issues concerning solar energy in distribution grids study, unrestricted renewable generating, such as residential solar (PV) systems, affects the characteristics of the load on distribution networks (Niitsoo et al. 2015). This study looked at how power quality measures in residential distribution networks are impacted by the harmonic content of PV generation. A more straightforward harmonics model for PVs was suggested after the outcomes of several modelling scenarios were analysed. The research discovered a modest increase in harmonic distortion in residential load current and voltage distortion when PVs were added to the substation's bus bar. Because it recorded and analysed harmonic current values at higher orders, the piece stood out (Li et al. 2021; Sevilla-Salcedo et al. 2021). The purpose of utilizing solar energy systems for the generation of electrical energy with the use of a solar photovoltaic (SPV) system is to satisfy the residential load requirements of a typical building. After finishing the design and economic analysis of the solar photovoltaic energy system for the different costs of the solar photovoltaic (SPV) module, a comparison was made between the grid power and the SPV system, both with and without storage. The results of the research-backed the use of the SPV system in residential buildings and demonstrated that the solar photovoltaic (SPV) system is an economically practical alternative for addressing India's rapidly expanding energy demand for home applications. The study was carried out by solar power ventures (SPV). In a method for visualizing and analysing smart-grid data created by synchro phasors with a visualization tool and a density-based clustering strategy. This method was described in the context of smart-grid data. Using a tool that is based on the circle representation in MATLAB, the real-time phasor data that were created by a smart-grid model that replicates a synchrophasor were shown. A density-based clustering approach was used to cluster the phasor data to detect contingency circumstances such as bad-data categorization, various fault types, and deviation on frequency, voltage, or current value deviations. This was done to enhance situational alertness and improve overall situational awareness. In this article, the data from an IEEE 14 bus system test-bed that was modelled in MATLAB/SIMULINK were utilized to provide system operators with assistance in doing various predictive analytics and making various selections (Aladi et al. 2018; Cannizzaro et al. 2021).

The study on the subject of the economic evaluation approach of solar power generation installed in ordinary homes set out to provide a GIS-based technique for assessing the financial benefits of putting photovoltaic panels on regular houses. To provide a concise summary of the results, we chose the following three points: (a) This method determined the profit and loss as well as the pay-back period to evaluate the installation of photovoltaic power generation by taking into account the price of the equipment, the solar battery module conversion efficiency, the subsidy, the price of the electricity purchase, the service life, and the rate for selling electricity. (b) This method was used to evaluate the installation of photovoltaic power generation. The proposed method of evaluation was put to the test in the prefecture of Kanagawa in Japan, where several different scenarios were developed. By using a solar battery module with a conversion efficiency of more than 15%, it could achieve a payback period that was less than the 20-year service life and anticipate making a profit throughout the vast majority of the territory. Kawasaki City and Ninomiya-machi are both excellent locations for the generation of solar electricity. It was required to take efforts to boost the subsidy and develop solar power production equipment in specified places in regions where subsidies are not supplied in appropriate quantities. These are the areas where the steps were taken.

The research investigated the significance of smart grids and carried out an in-depth feasibility study of the concept's potential use in Bangladesh. Both the characteristics of a smart grid and a conventional grid system were analysed and contrasted with one another. In addition to this, it investigated the best ways to integrate information and renewable energy technology into electricity distribution and transmission networks (Alam et al. 2006; Hossain et al. 2020). When it comes to optimum planning for stand-alone photovoltaic (SPV) systems, factors like dependability and economic performance need to be taken into consideration. (Ling and Liu 2013) devised a feasibility calculation approach to do just that. A sensitivity analysis of component costs and load profiles was conducted to illustrate the consequences of system uncertainty. Recent research on electrical energy storage devices geared at smart grid deployment has focused on the economic and environmental advantages of the smart grid (Biswas et al. 2013). The smart grid's better communications and control qualities would make it feasible to accommodate the potential usage of widely dispersed battery storage devices. The distributed microgrid control and applied demand response algorithm proposed in demand and response in smart grids for modern power systems (Raza et al. 2013) increase power system reliability. In this article, we looked at how microgrids use conditions to regulate demand and response. The microgrid TCP/IP setup that was recommended offered the shed a fighting chance against any potential disasters. (Abdel-aal et al. 2012) created a feasibility study for a solar energy stand-alone system (S.A.S.E.S.) to provide power to a small town. To provide a baseline for the investigation, consider that 5 kWh is needed to produce one cubic meter of hydrogen by electrolysis in 5 h. A solar module, water electrolyser, and fuel cell were the primary building blocks of the proposed system. The financial viability of the proposed system is examined. The projected hydrogen cost has been found by calculations. (Jamil et al. 2012) conducted a literature assessment on the topic of solar photovoltaic power generation's techno-economic viability. Three main categories were identified for literature categorization: design methods, the techno-economic viability of solar photovoltaic power production, and performance evaluations of various systems.

(Mujtaba et al. 2020) suggested as an alternative to investing in physical upgrades, distributed generation (DG) connection/injection opportunities may be expanded via the use of global active network management (ANM). The paper includes the results of an ANM feasibility study looking at potential congestion issues in the complete network. New connection standards and network access concepts are explored. It stressed how important it is to look at the problem from all angles to develop a solution that is both practical and efficient. In addition to the ANM feasibility study, further studies focused on adaptable access choices in real-world, limited-scale pilot projects. A study on the simulation and analysis of a multilevel converter topology for a solar photovoltaic (PV)-based grid-connected inverter (Adil et al. 2011) revealed that the utilization of PV-based systems for power production has increased significantly owing to the PV's renewable nature. This work introduced and analysed a multilayer line commutated inverter topology for improving the wave shape of the line current and consequently reducing the total harmonic distortion (THD) in a grid-tied line commutated inverter. The strategy worked when put into action and tested. The performance of the proposed topology was also shown to be far better than that of a conventional line commutated inverter. We see decreases in THD, losses, switching stress, and EMI.

3.8 Summary of Major Researches

The study's overarching objective is to determine whether solar radiation levels in Kalaburagi followed a pattern typical to other locations between 2004 and 2014. These first studies were predicated on data from 2004–2014, and since then, no more research has been conducted to evaluate the worldwide upward trend in surface solar radiation. An overview of these studies is shown in Table 3.3.

The collection of articles with the following search keywords results in only 35 significant articles.

TITLE-ABS-KEY (renewable AND energy AND + AND Indian AND smart AND cities)

3.9 Review of Solar Radiation Modelling Software's

Over the past five decades, many building energy simulation programmes and models have been created to analyse solar energy potential in buildings with significant progress being made, especially in the past two decades (Dunlop et al. 2003; Hofierka and Šúri 2002). These models and simulations are becoming ever more important with the increasing demand for electricity throughout the world, and in particular, the urban environment (Tito et al. 2016; Tudu et al. 2012). The following analysis is broken down into three parts, each of which contains a discussion of one of the many different kinds of solar radiation models: (1) Models of solar radiation based on GIS data, (2) models of solar radiation created using open source software, and (3) other software tools for modelling solar radiation This part of the article will

Table 3.3 Summary of significant research contribution

Authors	Title	Year	Citations
Krishankumar R., Pamucar D., Deveci M., Aggarwal M., Ravichandran K.S	Assessment of renewable energy sources for smart cities' demand satisfaction using multi-hesitant fuzzy linguistic-based choquet integral approach	2022	2
Rena, Yadav S., Patel S., Killedar D.J., Kumar S., Kumar R	Eco-innovations and sustainability in solid waste management: An Indian upfront in technological, organizational, start-ups and financial framework	2022	2
Dheeraja K., Padma Priya R., Ritika T	Optimal real-time pricing and sustainable load scheduling model for smart homes using Stackelberg game theory	2022	
Govindarajan H.K., L.S. G	Renewable energy for electricity use in India: Evidence from India's smart cities mission	2021	4
Pramanick S., Ahmad N., Jain D.K	Real-time electricity market (RTM) in India	2021	
Sivanand A., Devanathan C., Shankar E., PrasannaKumar P	Wear analysis of epoxy resin composites reinforced with seashell	2021	
Akhtar I., Kirmani S	Impact of sustainable design on India's smart cities development	2021	2
Sambhi S., Sambhi S., Bhadoria V.S	IoT-based optimized and secured ecosystem for energy Internet: The state-of-the-art	2021	1
Telang S., Chel A., Nafdey R., Kaushik G	Solar energy for sustainable development of a smart city	2021	2
Ganesh V., Senthilmurugan S., Ajay Krishna V.M., Ajit Ram R.R., Prabhu A	Smart grid-meters and communications-design, challenges, issues, opportunities and applications	2020	1
Praveen M., Sivakrishna Rao G.V	Implementation opportunities and challenges for renewable energy resources-based smart microgrids in India	2020	5
Bano F., Sehgal V., Tahseen M	Early stage design guidelines for net-zero-energy office buildings in tropical climate	2020	2
Singh M., Leena G	Forecasting of waste-to-energy system: A case study of Faridabad, India	2020	9
Suthar N	Estimating effect of electric vehicles on Indian grid system and identifying consequent opportunities	2020	

(continued)

Table 3.3 (continued)

Authors	Title	Year	Citations
Chakraborty S., Das S., Negi M	Hybrid microgrids for diesel consumption reduction in remote military bases of India	2020	1
Kulkarni A.R., Rajurkar S.S., Ballal M.S	Wide area monitoring system (WAMS): The foundation block of smartgrids for event analysis	2020	
Narasimalu S	Floating clean multi-energy systems towards driving blue economic growth	2020	1
Dixit A	Electric vehicle infrastructure market sustainable growth in Indian scenario, needs and suggestions	2019	3
Kumar D.S., Savier J.S	Impact analysis of distributed generation integration on distribution network considering smart grid scenario	2017	11
Sinha A.A., Shukla A., Prasad R.B	A review on CSP technologies with heat transfer fluids used in Indian power plants	2017	2
Santhiyakumari N., Shenbagapriya M., Hemalatha R	A novel approach in information and communication technology combined with traditional practices for smart villages	2017	4
Kumar R., Bhattacharya B., Kumar S.V.V.A., Chakraborty A., Bhowmick S.A., Sharma R., Sharma S	Harnessing non-conventional energy potential using space borne sensors	2017	
Sharma A.K., Dogra V.K	Preparation of papers - potential alternate energy resources for sustainability: A must need for a top pilgrimage city	2017	3
Gupta A., Mishra S., Bokde N., Kulat K	Need of smart water systems in India	2016	41
Banerjee R	Securing cities for energy needs	2015	
Das A., Balakrishnan V	Grid-connectivity of remote isolated islands-A proposition in Indian context	2012	2
Balijepalli V.S.K.M., Khaparde S.A	Smart and sustainable energy systems for developing countries: An Indian perspective	2011	7

come to a close with a table that provides a concise summary of the advantages and disadvantages of the solar software tools that have been investigated.

3.9.1 Open Source Solar Radiation Models

Presently, there are four most common open-source solar radiation models r.sun, ESP-r, Radiance, and EnergyPlus, which are all freely available on the Internet and widely utilized. R.sun, ESP-r, and Radiance are three programmes that can create 3D models of structures and compute solar radiation. EnergyPlus, on the other hand, cannot generate 3D models on its own and must be used in conjunction with other solar software. The popular Perez model is the default approach for calculating diffuse solar radiation in all four apps (Chow et al. 2003; Chow et al. 2005). All of these programmes are well-known, and they will be covered as (i) R.sun, (ii) ESP-r, (iii) Radiance, and (iv) EnergyPlus; (ii) R.sun, (ii) ESP-r, (iii) Radiance, and (iv) EnergyPlus.

(1) **R.sun**

The R.sun solar radiation model was developed by Hofierka and Suri, and it was included as a module in the GRASS open-source environment of the Geographical Resources Analysis Support System (GIS). The C programming language was used to build the module (2002). R.sun is a highly sophisticated and adaptive solar radiation model that is based on a complete approach for geographically and temporally distributed solar radiation data computation. R.sun was developed by (Hofierka and Šúri 2002). For diverse weather situations (overcast, clear sky), and horizontal or inclined surfaces, but not for vertical facades, it can compute all three components of solar radiation (Hofierka et al. 2014). R.sun is a raster-based tool that works with data that is geographically changeable (for inputs and outputs). It is very useful for modelling huge and complicated areas and may be used for long-term computations at various map sizes (Hofierka and Šúri 2002). R.sun has a shadowing method that makes it simple to analyse sky obstacles (shadows) cast by the surrounding environment, such as buildings and objects. It may also be used to calculate the spatio-temporal fluctuation of albedo as a result of solar radiation (Xiangdong et al. 2021). Using shell scripts, R.sun can calculate shadows in time interval increments of up to thirty minutes. R.sun is a powerful piece of software that may be used to evaluate shadows cast by horizontal and inclined surfaces. However, to use it, extensive programming skills are required. As a result of R.sun's development in an open-source GRASS GIS environment, its source code is licensed under a general public license, which makes it simple for anybody to use and paves the way for further development.

(2) **ESP-r**

The second significant open-source programme is called ESP-r, and it's an integrated building energy simulation software tool that's been used extensively for more than twenty-five years (Iqbal and Iqbal 2019). It was penned using the UNIX operating system, and it was distributed as open-source software after its completion. More software packages such as computer-aided design (CAD) for visualization, Energy Plus, and Radiance are also examples of programmes that ESP-r may be compatible with. Climate display and analysis, shading and insolation pattern computation, and 2D and 3D model builders are some of the key aspects of ESP-support r's modules (Kaur et al. 2020; Santos et al. 2014). Most crucially, ESP-r can simulate PV electricity generation potential while taking into consideration building facades and rooftops. It is, however, not particularly user-friendly, necessitates programming expertise, and appears to be too much of a research tool. ESP-r likewise looks to offer greater potential for research than for practical use.

(3) **Radiance**

Radiance is a lighting simulation programme that analyses the amount of global solar radiation and illumination reaching buildings, facades, and rooftops. Radiance is a backward ray-tracing technology created at the Lawrence Berkeley Laboratory in 1984. It is centred on the Perez diffuse solar radiation model, which is believed to be extremely accurate. The energy perspective for the active and passive solar heating system may be reliably quantified using this programme. It's a UNIX-based software that serves as STIMAP's rendering engine (spatio-temporal irradiation mapping). It produces exceptionally exact results for daylight modelling as a consequence of its foundation in the calculation of images of total annual irradiation based on hourly data from meteorological stations. Radiance is a method that provides accurate measurements of the quantity of sunlight that strikes roofs and facades. The fact that it is essentially a Radiance programme and is quite script-driven in terms of its lighting simulation capabilities is a drawback. The lighting simulation tool known as Radiance assesses the quantity of global solar radiation and illumination that is received by structures, including their facades and roofs. Radiance is a technique that was developed at the Lawrence Berkeley Laboratory in 1984 that traces rays in the opposite direction It is predicated on the Perez diffuse solar radiation model, which has a strong reputation for precision in its predictions. With the help of this tool, one can properly compute the energy potential of both active and passive solar heating systems. This software runs on UNIX and acts as the rendering engine for STIMAP (spatio-temporal irradiation mapping). It generates exceptionally exact results for daylight modelling as a consequence of its foundation in the calculation of images of total annual irradiation based on hourly data from meteorology and its use of these data. Radiance is a method that provides accurate measurements of the quantity of sunlight that strikes roofs and facades.

(4) **EnergyPlus**

EnergyPlus is a well-organized and modular software programme, and its primary function is that of a building energy modelling engine. The US Department of Energy was the one who developed it in the 1970s, and it makes use of the Perez algorithm. It only accepts text files as input and output, and it lacks a user-friendly graphical interface. Because EnergyPlus is considered to be a fairly accurate estimating tool, several solar radiation modelling software products enable data sharing with it. Its biggest drawback is that it is a command-line programme that only produces text files.

3.9.2 *Other Solar Radiation Modelling Software Tools*

Other prominent solar radiation modelling software tools covered in this area include (i) BSim, (ii) TRNSYS, (iii) Ecotect, (iv) RETScreen, and (v) HOMER, which do not exactly fall into any of the categories utilized in the scientific community.

(a) **BSim**

BSim is one of Denmark's most widely used simulation programmes, having been in use for more than two decades. SimView (a graphic editor), SimLight (daylight analysis), XSun (direct sunlight and shadowing analysis), SimPV (PV electricity potential), and SimDXF are the major modules of this user-friendly simulation suite that incorporates numerous distinct computer modules (allows for import from CAD programmes). SimPV can generate results on an hourly basis, allowing you to compare a PV system's anticipated hourly electrical production with other hourly power usage data. BSim can do a complete study of shadows cast by nearby structures and generate animation sequences for solar effects. It is gaining in popularity across the world and is useful for conducting extensive studies of building PV power-generating potential. Due to the SimPV module's reduced routines and algorithms, it is still considered a simple design tool.

(b) **TRNSYS**

TRNSYS is a modular-structured transient system simulation application that allows dynamic simulations for PV systems. It was created in the 1970s to solve complicated energy system challenges. TRNSYS is an excellent tool for undertaking in-depth assessments on any system whose behaviour is time-dependent. Calculation and modelling of solar radiation potential for PV systems are its key uses. TRNSYS is capable of producing and developing 3D geometric models, as well as supporting data transfers via tools like MATLAB and Simulink. Its biggest drawback is that 3D modelling software is not immediately integrated into the application; instead, users must run and download TRNSYS3d for Google Sketch Up, which is based on the open-source programme OpenStudio.

(c) **Ecotect Analysis**

Ecotect Analysis is a cutting-edge software application from Autodesk, the world's top provider of 2D and 3D design software. Ecotect is a sophisticated analytical tool that can calculate a building's whole solar energy capability in a complex metropolitan environment. Ecotect was developed by Ecotect. It is coupled to a powerful 3D modeller that is equipped with a variety of performance analysis tools that can manage the visualization, thermal performance, solar energy analysis, sun-path diagrams, shading, ventilation, environmental consequences, and cost of the simulation process. When compared to alternative solar radiation modelling tools, Ecotect offers several advantages, including the ability to provide practically rapid feedback at any point of the design process, interactive information displays, and the ability to make extremely aesthetically appealing drawings. It can also import and export data to other solar radiation analysis software applications like EnergyPlus, ESP-r, and Radiance, as well as a variety of formats for most CAD systems. Ecotect also has a strong scripting engine for triggering analytic tools and analysing findings, giving it a lot of flexibility. Most significantly, it is extremely user-friendly and gives highly accurate estimations of solar potential energy in a single package.

(d) **RETScreen**

RETScreen Clean Energy Project Analysis Programme is an energy decision-making software based on statistical analysis. It is one of the world's premier renewable energy decision-making applications from the Government of Canada makes it free to download. RETScreen is a spreadsheet-based programme that provides comprehensive energy and financial analysis for renewable energy technologies like wind and solar. It may examine energy output, life-cycle costs, and reductions in greenhouse gas emissions. RETScreen's primary drawback is that it can only create twelve-monthly power values for PV systems and cannot produce the full 8760 points of data that other hourly simulation models can.

(e) **HOMER**

The US National Renewable Energy Laboratory created the hybrid optimization model for electric renewables (HOMER), which is an energy modelling programme (NREL). It's a strong analytical tool that includes optimization and sensitivity analysis methods and is typically utilized as a cost–benefit analysis model. HOMER includes a wide range of features and technological options, allowing users to simulate the physical behaviour of a power system as well as its life-cycle costs. It can estimate hourly power output for solar PV applications as well as other renewable energy sources depending on input parameters. HOMER calculates the energy balance for each of the 8760 h in a year to replicate the PV system. HOMER's ability to synthesize 8760 h electricity values from twelve monthly average daily radiation readings is one of its primary advantages. This is accomplished by HOMER using an algorithm created particularly for solar simulation design work. It creates hourly solar irradiation data sets using simply the twelve-monthly mean of daily occurrences known as the global horizontal irradiance (GHI) and the location's latitude.

3.9.3 *Summary of the Softwares*

In today's environment, there are a plethora of solar modelling software packages to choose from. This review detailed the most often discussed software. This programme has its own set of capabilities, each with its own set of strengths and drawbacks. Not all programmes are used globally, and some appear in the literature more frequently than others. The capabilities of this various software are summarized in Table 3.4. The HOMER and RETScreen were utilized in this study for a variety of studies, including feasibility and spatio-temporal analysis.

3.10 Summary

The section provides an overview of modelling, beginning with the collecting of data and continuing through the assessment of the final proposed models. It was discussed how solar radiation is measured, and a short history of the techniques that are now accepted and widely used was also presented. There was also a discussion of other meteorological characteristics that are presently significant to this investigation. We also spoke about the possibility of error in measurements, as well as other kinds of faults associated with them. In the research that was done, a variety of data quality control programmes were investigated. Depending on the kind of data that was fed into the models, two primary categories of solar radiation simulations were found to exist. They were then broken down into subgroups based on the regression link between the different parameters that were included in each category. In addition to that, there was a concise introduction to satellite models and models of atmospheric transmittance (including the spectral type). In conclusion, a wide variety of statistical indicators that have been employed in the modelling literature have been investigated and commented on. To summarize, the precision of the measurements is of the utmost importance when it comes to the construction of models. These models, after they have been validated and tested, should exhibit adequate estimate accuracy within the given uncertainty of the observed data.

Advantages and Limitations of Solar Energy

Solar energy offers an opportunity for augmented provide energy services with reduced emissions that are either zero or very close to zero. There is no other renewable energy source that can compete with the sun's abundant supply of energy, and solar power is the only one that can. Every new technology comes with both advantages and disadvantages. Solar insolation and atmospheric conditions have a significant impact on the amount of energy that can be harvested from the sun.

(a) *Advantages*

1. It has a large supply of energy derived from renewable sources.
2. This technology is present everywhere, and it has the capability of being acquired and transformed consistently.

Table 3.4 Summary of the solar insolation modelling tools

Software tools	Strength	Weakness	3D data support	Expertise level	Open source/commercial	Time sensitivity of estimate
ESP.r	– Accurate calculation algorithms – Compatible with other solar software	– Modelling knowledge is required and uneasy to use – Bad user interface	Yes	High	Yes	Hourly
R.sun	– Capable of beautiful maps – Best for large–scale analysis – Capable of handling complex terrain maps	– Not suitable for small–scale analysis	Yes	High	Yes	Capable of generating up to 30 m
Radiance	– Able to quantify solar potential – can generate 3D images	Good for light simulation and rendering of images	Yes	High	Yes	Could not traced
Energy Plus	– Structured software – Accurate calculations	– No graphical user interface – Command line based – Only text-based inputs/outputs	Yes	High	Yes	Hourly

(continued)

Table 3.4 (continued)

Software tools	Strength	Weakness	3D data support	Expertise level	Open source/commercial	Time sensitivity of estimate
TRNSYS	– Can perform detailed calculations – Modular structure software	– Not capable of performing analysis on multiple buildings	Yes	Moderate	Yes	Hourly
RETScreen	– Spreadsheet approach – User friendly	– Cannot generate hourly electricity potential	No	Low	Yes	Monthly
HOMER	– Can generate accurate predictions for hourly PV estimations	– Does not have a visually appealing interface	No	Moderate	No	Hourly

Source Adapted from (Chow 2012)

3. This technology does not contribute to global warming in any way since it does not release any greenhouse gases.
4. Because the process of creating energy does not include any moving parts, the technology in question does not produce any audible noise.
5. This technology requires less maintenance since it does not have any moving components.
6. It is possible to arrange it in a modular form, and then extend it over time.
7. One of the most workable strategies for bringing electricity to distant rural areas is to construct it in areas where there is a strong energy demand and then expand it in a modular fashion. This strategy has many advantages.

(b) *Limitations*

1. The conversion efficiency levels of light to energy vary from 10 to 17% depending on the technology, since the technology is still in its early phases and is still in its developmental stages.
2. The initial expenditure required to use this technology is substantial. The implementation of various subsidy programmes by the government has allowed the technology to continue existing.
3. The use of solar electricity is restricted to the hours when the sun is up. According to the vast majority of load profiles, the peak load occurs in the evening or sometime during the night. Because of this, it is necessary to make use of pricey storage devices such as batteries, which need to be replaced every three and five years. The cost of the Battery may often contribute to between 30 and 40 per cent of the entire cost of the system.
4. As a result of the low levels of efficiency, the quantity of space that is necessary is rather substantial. For instance, with the technology that is available today, the construction of a solar PV power plant with a capacity of 1 MW would need between 6 and 9 acres of land.

Affordable and dependable energy has helped people live longer, healthier lives. Billions still lack clean cooking fuel or power. As more people acquire access to energy and improve their living conditions, the global middle class will rise. More dwellings, transportation, electricity, consumer goods, and energy are needed. Cities account for more than 70% of greenhouse gas emissions. Cities can lead the move to sustainable energy since they emit so much GHG. New data illustrates how cities are making this transformation by signing arrangements to switch their municipal operations and, in some instances, their whole neighbourhoods to renewable energy. To accommodate 2.5 billion urban dwellers by 2050, cities must be low-carbon, resilient, and liveable. Cities generate the most anthropogenic greenhouse gases and are most sensitive to climate change and catastrophic weather. We research methods for constructing sustainable energy systems by lowering energy usage in buildings and transportation and offering decentralized, renewable energy sources. Increasing power density can help city-integrated renewable energy meet the energy needs of increasing metropolitan regions. Innovation must address economic, technological, behavioural, and political issues to increase urban sustainability.

References

Abdel-aal H, Bassyouni M, Abdelkreem M, Abdel-hamid S, Zohdy K (2012) Feasibility study for a solar-energy stand-alone system 204–206

Abnavi MD, Mohammadshafie N, Rosen MA, Dabbaghian A, Fazelpour F (2019) Techno-economic feasibility analysis of stand-alone hybrid wind/photovoltaic/diesel/battery system for the electrification of remote rural areas: case study Persian Gulf coast-Iran. Environ Prog Sustain Energy 38(5):1–15. https://doi.org/10.1002/ep.13172

Adelaja S, Shaw J, Beyea W, Charles McKeown JD (2010) Renewable energy potential on brownfield sites: a case study of Michigan. Energy Policy 38(11):7021–7030. https://doi.org/10.1016/j.enpol.2010.07.021

Adelaja S, Shaw J, Beyea W, Charles McKeown JD, Agugiaro G, Nex F, Remondino F, Filippi R. De, Droghetti S, Furlanello C, Foundation BK, Stevanato G, Filippi R. De, Furlanello C, Kessler FB, Modelling U, Unit E, Al-enezi FQ, Sykulski JK (2011) Potential and prospects of solar energy in Uttara Kannada, District of Karnataka State, India. Solar Energy 02(2):1986–2000. https://doi.org/10.1080/00908319708908903. www.imagingnotes.com

Agugiaro G, Remondino F, Stevanato G, Filippi RD, Furlanello C, Kessler FB, Modelling U, Unit E (2005) Estimation of solar radiation on building roofs in mountainous areas. Int Arch Photogrammetry Remote Sens Spat Inf Sci 38:155–160

Ahmad MJ, Tiwari GN (2010) Solar radiation models—review. Int J Energy Environ 1(5):861–870

Ahmad F, Goparaju L (2018) A geospatial analysis of climate variability and its impact on forest fire: a case study in Orissa state of India. Spat Inf Res 26(6):587–598. https://doi.org/10.1007/s41324-018-0197-8

Ajayi O, Ohijeagbon O, Ajanaku K, Aasa S, Omotosho O (2016) Techno-economic assessment of renewable hybrid systems for rural electrification and distributed generation in selected sites across Nigeria. J Afr Res Bus Technol 2015:1–31. https://doi.org/10.5171/2015.124767

Al-enezi FQ, Sykulski JK, Ahmed NA (2011) Energy procedia visibility and potential of solar energy on horizontal surface at Kuwait area. Energy Procedia 12:862–872. https://doi.org/10.1016/j.egypro.2011.10.114

Aladi HB, Saha S, Kurian A, Basu A (2018) Predictive analytics for safer smart cities. In: Proceedings of the 2017 international conference on smart technology for smart nation, SmartTechCon 2017, pp 1010–1017. https://doi.org/10.1109/SmartTechCon.2017.8358523

Alam MJB, Alam MJB, Rahman MH, Khan SK, Munna GM (2006) Unplanned urbanization: assessment through calculation of environmental degradation index. Int J Environ Sci Technol 3(2):119–130. https://doi.org/10.1007/BF03325915

Alijani Z, Hosseinali F, Biswas A (2020) Spatio-temporal evolution of agricultural land use change drivers: a case study from Chalous region, Iran. J Environ Manage 262(110326). https://doi.org/10.1016/j.jenvman.2020.110326

Amani M, Ghorbanian A, Ahmadi SA, Kakooei M, Moghimi A, Mirmazloumi SM, Moghaddam SHA, Mahdavi S, Ghahremanloo M, Parsian S, Wu Q, Brisco B (2020) Google Earth engine cloud computing platform for remote sensing big data applications: a comprehensive review. IEEE J Sel Top Appl Earth Observations Remote Sens 13:5326–5350. https://doi.org/10.1109/JSTARS.2020.3021052

Anand P, Bath S, Rizwan M (2017) Design of solar-biomass-biogas based hybrid system for rural electrification with environmental benefits. Int J Recent Innov Trends Comput Commun 5(6):450–456. http://www.ijritcc.org/download/browse/Volume_5_Issues/June_17_Volume_5_Issue_6/1497679692_17-06-2017.pdf

Andalibi L, Ghorbani A, Moameri M, Hazbavi Z, Nothdurft A, Jafari R, Dadjou F (2021) Leaf area index variations in ecoregions of Ardabil Province, Iran. Remote Sens 13(15). https://doi.org/10.3390/rs13152879

Anderson MC, Norman JM, Kustas WP, Houborg R, Starks PJ, Agam N (2008) A thermal-based remote sensing technique for routine mapping of land-surface carbon, water and energy fluxes from field to regional scales. Remote Sens Environ 112(12):4227–4241

Anwar K, Deshmukh S (2018) Assessment and mapping of solar energy potential using artificial neural network and GIS technology in the Southern part of India. Int J Renew Energy Res 8(2):1–12

Ashrafi Goudarzi S, Fazelpour FB, Gharehpetian G, Rosen MA (2019) Techno-economic assessment of hybrid renewable resources for a residential building in tehran. Environmen Prog Sustain Energy 38(5):1–14. https://doi.org/10.1002/ep.13209

Azapagic A (2004) Developing a framework for sustainable development indicators for the mining and minerals industry. J Clean Prod 12(6):639–662. https://doi.org/10.1016/S0959-6526(03)00075-1

Baniasad Askari I, Baniasad Askari L, Kaykhah MM, Baniasad Askari H (2014) Optimisation and techno-economic feasibility analysis of hybrid (photovoltaic/wind/fuel cell) energy systems in Kerman, Iran; considering the effects of electrical load and energy storage technology. Int J Sustain Energ 33(3):635–649. https://doi.org/10.1080/14786451.2013.769991

Baniya B, Techato K, Ghimire SK, Chhipi-shrestha G (2018) A Review of green roofs to mitigate urban heat Island and Kathmandu Valley in Nepal. Appl Ecol Environ Sci 6(4):137–152. https://doi.org/10.12691/aees-6-4-5

Basir R, Aziz N, Ahmad SS, Wahid A (2013) Satellite remote sensing for identification of solar potential sites in Pakistan. Int J Basic Appl Sc 2(2):200–209

Bawah U, Addoweesh KE, Eltamaly AM (2013) Comparative study of economic viability of rural electrification using renewable energy resources versus diesel generator option in Saudi Arabia. J Renew Sustain Energy 5(4). https://doi.org/10.1063/1.4812646

Biswas M, Azim S, Saha TK, Zobayer U (2013) Towards implementation of smart grid : an updated review on electrical energy storage systems 122–132. https://doi.org/10.4236/sgre.2013.41015

Bois B, Wald L, Pieri P, Van Leeuwen C, Commagnac L, Chery P, Christen M, Gaudillère JP, Saur E (2008) Estimating spatial and temporal variations in solar radiation within Bordeaux winegrowing region using remotely sensed data. J Int Sci Vigne Du Vin 42(1):15–25. http://hal.archives-ouvertes.fr/hal-00363698

Bojanowski JS (2013) Quantifying solar radiation at the earth surface with meteorological and satellite data.

Cannizzaro D, Aliberti A, Bottaccioli L, Macii E, Acquaviva A, Patti E (2021) Solar radiation forecasting based on convolutional neural network and ensemble learning. Expert Syst Appl 181. https://doi.org/10.1016/j.eswa.2021.115167

Chow A (2012) Modeling urban solar energy with high spatiotemporal resolution

Cooper CM, Troutman JP, Awal R, Habibi H, Fares A (2022) Climate change-induced variations in blue and green water usage in U.S. urban agriculture. J Cleaner Prod 348. https://doi.org/10.1016/j.jclepro.2022.131326

Datta A, Karakoti I (2010) Solar resource assessment using GIS and remote sensing techniques 9560272741(M):1–20. ESRI India User Conference

Dey S, Tripathi SN, Singh RP, Holben BN (2005) Seasonal variability of the aerosol parameters over Kanpur, an urban site in Indo-Gangetic basin. Adv Space Res 36(5):778–782. https://doi.org/10.1016/j.asr.2005.06.040

Dheeraja K, Padma Priya R, Ritika T (2022) Optimal real-time pricing and sustainable load scheduling model for smart homes using stackelberg game theory. In Lecture notes on data engineering and communications technologies, vol 99. https://doi.org/10.1007/978-981-16-7182-1_22

Dixit A (2019) Electric vehicle infrastructure market sustainable growth in indian scenario, needs and suggestions. Int J Innovative Technol Exploring Eng 8(11 Special):242–245. https://doi.org/10.35940/ijitee.K1048.09811S19

Dovì VG, Friedler F, Huisingh D, Klemeš JJ (2009) Cleaner energy for sustainable future. J Clean Prod. https://doi.org/10.1016/j.jclepro.2009.02.001

Dubayah R, Rich PM (1996) GIS-based solar radiation modeling. GIS Environ Model 0:129–134

Dunlop ED, Šúri M, Huld TA (2003) Photovoltaic potential assessment to support renewable energies growth in 10 EU candidate Countries, pp 3–7

Dwevedi R, Krishna V, Kumar A (2018) Environment and big data: role in smart cities of India. Resources 7(4):64. https://doi.org/10.3390/resources7040064

El-Zeiny AM, Effat HA (2017) Environmental monitoring of spatiotemporal change in land use/land cover and its impact on land surface temperature in El-fayoum governorate. Egypt Remote Sens Appl Soc Environ 8:266–277. https://doi.org/10.1016/j.rsase.2017.10.003

Ezzine H, Bouziane A, Ouazar D (2014) Seasonal comparisons of meteorological and agricultural droughtindices in Morocco using open short time-series data. Int J Appl Earth Obs Geoinf 26(1):36–48. https://doi.org/10.1016/j.jag.2013.05.005

Fargas DCJ, Narciso GAM, Blanco AC (2021) Monitoring and assessment of agri-urban land conversion using multi-sensor remote sensing and GIS techniques. In: Paparoditis NLFYMYJS-SAZHLXOBSUHESMZJPAWLLRYMDKAOAHMFFS, Mallet C (eds) ISPRS annals of the photogrammetry, remote sensing and spatial information sciences, vol 5, Issue 3. Copernicus GmbH, pp 117–124. https://doi.org/10.5194/isprs-annals-V-3-2021-117-2021

Fath K, Stengel J, Sprenger W, Wilson HR, Schultmann F, Kuhn TE (2015) A method for predicting the economic potential of (building-integrated) photovoltaics in urban areas based on hourly radiance simulations. Sol Energy 116:357–370. https://doi.org/10.1016/j.solener.2015.03.023

Gambhir A, Dixit S, Toro V, Singh V, Ramachandra TV, Jain R, Krishnadas G, Nayak JK, Prajapati JA, Noel, Cressie, Christopher KW, Lean JL, Sayigh AAM, Pareek A, Studios B, Commission CE, National T, Plan A, Change C, Nehru J, … Observatory N (2010) A Guide to PV System Design and Installation. In: Solar Energy, vol. 44, Issue 5. Academic Press. https://doi.org/10.1002/wcc.018

Gevelt van T, Holmes J (2015) Electricity for off-grid villages : an overview of the current technical report 2 (Issue February).

Gils HC, Scholz Y, Pregger T, Luca de Tena D, Heide D (2017) Integrated modelling of variable renewable energy-based power supply in Europe. Energy 123:173–188. https://doi.org/10.1016/j.energy.2017.01.115

Guleria A, Gupta SK, Gupta I, Swami D, Shukla DP (2019) Understanding the spatial and temporal dependence of the migration of conservative contaminant plume in urban groundwater environment in Panchkula region, Haryana, India. Groundw Sustain Dev 8:93–103. https://doi.org/10.1016/j.gsd.2018.10.003

Hall P, Zidar M (1993) Use of geographic information systems and models on personal computer and workstation platforms, in the development of a wellhead protection program, Salinas Valley, California. In: Proceedings of the symposium on engineering hydrology, pp 377–382. https://www.scopus.com/inward/record.uri?eid=2-s2.0-0027224345&partnerID=40&md5=a8f1c907e8666e9c28a35496c3b7efce

Hasan SH, Al-Hameedawi ANM, Ismael HS (2021) Using Google Earth Engine development environment for remote sensing image analysis, Al Shuwija marsh case study. In: Obaid AJSMSSSS, Albermany S (eds) J Phys: Conf Ser 1973(1). IOP Publishing Ltd. https://doi.org/10.1088/1742-6596/1973/1/012192

Hassan QK, Rahman KM, Haque AS, Ali A (2011) Solar energy modelling over a residential community in the City of Calgary, Alberta, Canada. Int J Photoenergy. https://doi.org/10.1155/2011/216519

Hochwimmer B (2009) GIS-based modeling of Solar Radiation (Issue July)

Hofierka J, Cebecauer T (2008) Spatially distributed assessment of solar resources for energy applications in Slovakia. In: Acta Facultatis Studiorum …, pp 97–114. http://web.science.upjs.sk/hofierka/projekty/Hofierka_Cebecauer_Folia12.pdf

Hofierka J, Kaňuk J, Gallay, M (2014) The spatial distribution of photovoltaic power plants in relation to solar resource potential: the case of the Czech Republic and Slovakia. Moravian Geogr Rep 22(2011). http://www.degruyter.com/view/j/mgr.2014.22.issue-2/mgr-2014-0009/mgr-2014-0009.xml

Hofierka J, Šúri M (2002) The solar radiation model for open source GIS : implementation and applications. In: Open source GIS—GRASS users conference, pp 11–13

Hossain R, Ahmed AJ, Islam SMKN, Saha N, Debnath P, Kouzani AZ, Mahmud MAP (2020) New Design of solar photovoltaic and thermal hybrid system for performance improvement of solar photovoltaic. Int J Photoenergy 2020(1). https://doi.org/10.1155/2020/8825489

Huang J, Mo Z, Liu H, Chen Z, Li H, Zhang D, Mu Y, Yang J, Liang G, Wang D (2021) Analysis on the causes of PM2.5 pollution in Nanning City in february 2020. Huanjing Kexue Xuebao/Acta Sci Circumstantiae 41(4):1173–1183. https://doi.org/10.13671/j.hjkxxb.2020.0445

Huber M, Dimkova D, Hamacher T (2014) Integration of wind and solar power in Europe: assessment of flexibility requirements. Energy 69:236–246. https://doi.org/10.1016/j.energy.2014.02.109

Iqbal A, Iqbal T (2019) Design and analysis of a stand-alone PV system for a rural house in Pakistan. Int J Photoenergy 2019:8

Islam MR (2012) Study of micro grid safety & protection strategies with control system infrastructures. Smart Grid Renew Energy 03(01):1–9. https://doi.org/10.4236/sgre.2012.31001

Jamil M, Kirmani S, Rizwan M (2012) Techno-economic feasibility analysis of solar photovoltaic power generation : a review 266–274

Ji H, Li X, Wei X, Liu W, Zhang L, Wang L (2020) Mapping 10-m resolution rural settlements using multi-source remote sensing datasets with the google earth engine platform. Remote Sens 12(17):1–23. https://doi.org/10.3390/rs12172832

Jiang F, Xie H, Ellen O (2018) Hybrid energy system with optimized storage for improvement of sustainability in a small town. Sustainability (switzerland) 10(6):1–16. https://doi.org/10.3390/su10062034

Jochem A, Wichmann V, Hofle B (2010) Large area point cloud based solar radiation modeling. In: Hamburger .... http://koenigstuhl.geog.uni-heidelberg.de/publications/2010/Hoefle/Jochem_et_al_2010_Solar_PointCloud_SAGA_Meeting.pdf

Joyi O, Utanga MM, Dindi OO, Ynekulu JA, Ahman EAB (2015) Hyper-temporal remote sensing for scaling between spectral indices and flux tower measurements. Appl Ecol Environ Res 13(2):465–487. https://doi.org/10.15666/aeer/1302

Kabisch N, Alonso L, Dadvand P, van den Bosch M (2019) Urban natural environments and motor development in early life. Environ Res 179. https://doi.org/10.1016/j.envres.2019.108774

Kamau-Devers K, Yanez VR, Medina Peralta VW, Miller SA (2021) Using internal micro-scale architectures from additive manufacturing to increase material efficiency. J Cleaner Prod 291. https://doi.org/10.1016/j.jclepro.2021.125799

Kandekar VU, Pande CB, Rajesh J, Atre AA, Gorantiwar SD, Kadam SA, Gavit B (2021) Surface water dynamics analysis based on sentinel imagery and Google Earth engine platform: a case study of Jayakwadi dam. Sustain Water Resour Manag 7(3). https://doi.org/10.1007/s40899-021-00527-7

Kannan D, Moazzeni S, Darmian SM, Afrasiabi A (2021) A hybrid approach based on MCDM methods and Monte Carlo simulation for sustainable evaluation of potential solar sites in east of Iran. J Cleaner Prod 279. https://doi.org/10.1016/j.jclepro.2020.122368

Kara F, Keçeli A (2017) Impact of rapid urbanisation on land cover in Istanbul Province. Spat Inf Res 25(2):293–302. https://doi.org/10.1007/s41324-017-0100-z

Kaur H, Garg P (2019) Urban sustainability assessment tools: a review. J Clean Prod 210:146–158. https://doi.org/10.1016/j.jclepro.2018.11.009

Kaur M, Dhundhara S, Verma YP, Chauhan S (2020) Techno-economic analysis of photovoltaic-biomass-based microgrid system for reliable rural electrification. Int Trans Electr Energy Syst 30(5):1–20. https://doi.org/10.1002/2050-7038.12347

Kazem HA, Khatib T, Sopian K, Elmenreich W (2014) Performance and feasibility assessment of a 1.4 kW roof top grid-connected photovoltaic power system under desertic weather conditions. Energy Build 82:123–129. https://doi.org/10.1016/j.enbuild.2014.06.048

Kim J, Kwan MP (2021) Assessment of sociodemographic disparities in environmental exposure might be erroneous due to neighborhood effect averaging: implications for environmental inequality research. Environ Res 195. https://doi.org/10.1016/j.envres.2020.110519

Kulkarni S, Banerjee R (2010). Renewable energy mapping in Maharashtra, India using GIS. In: World renewable energy congress 2011-Sweden, pp 3177–3184

Kumar D (2017) Surface temperature variability analysis of an urban area using landsat ETM + thermal images. Asian Geogr. https://doi.org/10.1080/10225706.2017.1322992

Kumar D (2019a) Mapping solar energy potential of southern India through geospatial technology. Geocarto Int 34(13):1477–1495. https://doi.org/10.1080/10106049.2018.1494759

Kumar D (2019b) Hyper-temporal variability analysis of solar insolation with respect to local seasons. Remote Sens Appl Soc Environ 15:100241. https://doi.org/10.1016/j.rsase.2019.100241

Kumar L, Skidmore AK, Knowles E (1997) Modelling topographic variation in solar radiation in a GIS environment. Int J Geogr Inf Sci 11(5):475–497. https://doi.org/10.1080/136588197242266

Landscape A, May UP (2016) Quantifying land surface temperature change from LISA clusters : an alternative approach to identifying urban land use transformation. Landscape and Urban Plann 153(May):51–65. https://doi.org/10.1016/j.landurbplan.2016.05.001

Laur AK (2014). Modelling the spatio-temporal variability of solar radiation on buildings: a case study of Lewis hall (Issue May).

Le HM, Corzo G, Medina V, Diaz V, Nguyen BL, Solomatine DP (2018) A comparison of spatial-temporal scale between multiscalar drought indices in the south central region of vietnam. In Spatiotemporal analysis of extreme hydrological events. Elsevier Inc. https://doi.org/10.1016/B978-0-12-811689-0.00007-0

Li P, Zhang Y, Zhang Y, Zhang K (2021) Prediction of electric bus energy consumption with stochastic speed profile generation modelling and data driven method based on real-world big data. Appl Energy 298. https://doi.org/10.1016/j.apenergy.2021.117204

Li S, Karava P, Currie S, Lin WE, Savory E (2014) Energy modeling of photovoltaic thermal systems with corrugated unglazed transpired solar collectors—Part 1: model development and validation. Sol Energy 102:282–296. https://doi.org/10.1016/j.solener.2013.12.040

Lian-tong Z (2011) Interdecadal variability in surface solar radiation over Northwest China and its possible cause. Atmos Oceanic Sci Lett 4(2):103–108

Ling J-M, Liu P-H (2013) Economic and feasibility analysis for stand–alone solar photovoltaic generation system. Energy Power Eng 05(04):356–362. https://doi.org/10.4236/epe.2013.54B069

Luna N, Medina J (2021) Implementation and assessment of temperature change due to volcanic eruptions using google earth engine: Galeras volcano nariño-colombia [Implementación y evaluación del cambio de temperatura por erupciones volcánicas usando google earth engine: Volcán ga. RISTI - Revista Iberica de Sistemas e Tecnologias de Informacao 2021(E41):437–447. https://www.scopus.com/inward/record.uri?eid=2-s2.0-85105521391&partnerID=40&md5=20ed277b212774de0e1464fe98429f4d

Luqman M, Ahmad SR, Khan S, Ahmad U, Raza A (2015) Estimation of solar energy potential from rooftop of punjab government servants cooperative housing society lahore using GIS. Sci Res Publishing 6(May):128–139

MacKerron G, Mourato S (2013) Happiness is greater in natural environments. Glob Environ Chang 23(5):992–1000. https://doi.org/10.1016/j.gloenvcha.2013.03.010

Marques-Perez I, Guaita-Pradas I, Gallego A, Segura B (2020a) Territorial planning for photovoltaic power plants using an outranking approach and GIS. J Cleaner Prod 257. https://doi.org/10.1016/j.jclepro.2020.120602

Meisen P, Quéneudec E, Yuan M, Nara A, Bothwell J, Ramirez L, Zink R, Dorner W, Stoeglehner G, Sharma P, Harinarayana T, Journal E, Studies E, Vol M, Division ES, Jacoby HD, Supervisor T, Marks DH, Systems E, … Ramkrishna B (2006) Overview of renewable energy potential of India. Model Earth Syst Environ 2(October):1–20. https://doi.org/10.1016/j.compenvurbsys.2015.03.002

Mohajerani A, Bakaric J, Jeffrey-Bailey T (2017) The urban heat island effect, its causes, and mitigation, with reference to the thermal properties of asphalt concrete. J Environ Manage 197:522–538. https://doi.org/10.1016/j.jenvman.2017.03.095

Molina MG, Espejo EJ (2014) Modeling and simulation of grid-connected photovoltaic energy conversion systems. Int J Hydrogen Energy 39(16):8702–8707. https://doi.org/10.1016/j.ijhydene.2013.12.048

Monedero J, Garcia J, Dobon F, Yanes MA, Hernandez F (2007) Calculation of PV potential maps in the canary islands. In: 22nd European photovoltaic solar energy conference and exhibition, 3127. c:/pdflib/00021911.pdf

Moutinho V, Madaleno M (2021) Assessing eco-efficiency in Asian and African countries using stochastic frontier analysis. Energies, 14(4). https://doi.org/10.3390/en14041168

Mujtaba G, Rashid Z, Umer F, Baloch SK, Hussain GA, Haider MU (2020) Implementation of distributed generation with solar plants in a 132 kV grid station at Layyah using ETAP. Int J Photoenergy. https://doi.org/10.1155/2020/6574659

Mulugetta Y, Drake F (1996) Assessment of solar and wind energy resources in Ethiopia. II. wind energy. Solar Energy 57(4):323–334. https://doi.org/10.1016/S0038-092X(96)00074-6

Myers DR (2005a) Solar radiation modeling and measurements for renewable energy applications: data and model quality. Energy 30(9). https://doi.org/10.1016/j.energy.2004.04.034

Myers DR (2005b) Solar radiation modeling and measurements for renewable energy applications: data and model quality. Energy, 30(9 SPEC. ISS.):1517–1531. https://doi.org/10.1016/j.energy.2004.04.034

Nandy S, Saranya M, Srinet R (2021) Spatio-temporal variability of water use efficiency and its drivers in major forest formations in India. Remote Sens Environ. https://doi.org/10.1016/j.rse.2021.112791

Negi S, Mathew L (2014) Hybrid renewable energy system: a review. Int J Electron Electr Eng 7(5):535–542. http://www.irphouse.com

Niitsoo J, Jarkovoi M, Taklaja P, Klüss J, Palu I (2015) Power quality issues concerning photovoltaic generation in distribution grids 148–163.

Nixon JD, Dey PK, Davies PA (2013) Design of a novel solar thermal collector using a multi-criteria decision-making methodology. J Clean Prod 59:150–159. https://doi.org/10.1016/j.jclepro.2013.06.027

Noh C, Kim I, Jang W, Kim C (2015) Recent trends in renewable energy resources for power generation in the republic of Korea. Resources 4(1):751–764. https://doi.org/10.3390/resources4040751

Nolasco D, Lima RN, Hernández PA, Pérez NM (2008) Non-controlled biogenic emissions to the atmosphere from Lazareto landfill, Tenerife, Canary Islands. Environ Sci Pollut Res 15(1):51–60. https://doi.org/10.1065/espr2007.02.392

Okwanya I, Alhassan A, Migap JP, Adeka SS (2021) Evaluating renewable energy choices among rural communities in Nigeria. An insight for energy policy. Int J Energy Sect Manage 15(1):157–172. https://doi.org/10.1108/IJESM-12-2019-0001

Olatomiwa L, Mekhilef S, Huda ASN, Ohunakin OS (2015) Economic evaluation of hybrid energy systems for rural electrification in six geo-political zones of Nigeria. Renew Energy 83:435–446. https://doi.org/10.1016/j.renene.2015.04.057

Ostapowicz K (2009). Solar energy modelling with ArcGIS.

Pandey P, Kumar D, Prakash A, Masih J, Singh M, Kumar S, Jain VK, Kumar K (2012) A study of urban heat island and its association with particulate matter during winter months over Delhi. Sci Total Environ 414:494–507. https://doi.org/10.1016/j.scitotenv.2011.10.043

Panidi E, Rykin I, Tsepelev V (2020) Satellite-based estimation of growing season framing dates, weather instability aspect. Arab J Geosci 13(10). https://doi.org/10.1007/s12517-020-05409-4

Perez R, David M, Hoff TE, Jamaly M, Kivalov S, Kleissl J, Lauret P, Perez M (2016) Spatial and temporal variability of solar energy. Found Trends® Renew Energy 1(1):1–44. https://doi.org/10.1561/2700000006

Perez R, Hoff TE, Schlemmer J (2011) Short-term irradiance variability. In: American solar energy society—proceeding of the ASES annual conference. Raleigh, NC 8

Perez, R., Schlemmer, J., Renne, D., Cowlin, S., George, R., & Bandyopadhyay, B. (2009). Validation of the suny satellite model in a meteosat environment. In: Proceeding of American solar energy society annual conference, Dec 2002. http://www.asrc.albany.edu/people/faculty/perez/publications/SolarResourceAssessmentandModeling/PapersonResourceAssessmentandSatellites/2009-validation-of-the-suny-model-in-a-meteosat-environment.pdf

Quader MA, Ahmed S, Raja Ghazilla RA, Ahmed S, Dahari M (2016) Evaluation of criteria for CO2 capture and storage in the iron and steel industry using the 2-tuple DEMATEL technique. J Clean Prod 120:207–220. https://doi.org/10.1016/j.jclepro.2015.10.056

Ramachandra TV, Subramanian, DK (1997) Potential and prospects of solar energy in Uttara Kannada, district of Karnataka State, India. Energy Sources 19(9):945–988. https://doi.org/10.1080/00908319708908903

Ramachandra TV (2007) Solar energy potential assessment using GIS. Energ Educ Sci Technol 18(2):101–114

Ramachandra TV, Kumar U (2009) Land surface temperature with land cover dynamics: multi-resolution, spatio-temporal data analysis of greater Bangalore, India. Int J Geoinf 5(3):43–53. https://www.scopus.com/inward/record.uri?eid=2-s2.0-79959761084&partnerID=40&md5=439f86427152125e941d7b6c3bf8f767

Ramachandra TV, Setturu B, Rajan KS, Subash Chandran MD (2017) Modelling the forest transition in Central Western Ghats India. Spat Inf Res 25(1):117–130. https://doi.org/10.1007/s41324-017-0084-8

Rauf S, Kalair AR, Khan N (2020) Variable load demand scheme for hybrid AC/DC nanogrid. Int J Photoenergy. https://doi.org/10.1155/2020/3646423

Raza MQ, Haider MU, Ali SM, Rashid MZ, Sharif F (2013) Demand and response in smart grids for modern power system 133–136

Rehman S, Ghori SG (2000) Spatial estimation of global solar radiation using geostatistics. Renew Energy 21(3–4):583–605. https://doi.org/10.1016/S0960-1481(00)00078-1

Ren S, Miles ES, Jia L, Menenti M, Kneib M, Buri P, McCarthy MJ, Shaw TE, Yang W, Pellicciotti F (2021) Anisotropy parameterization development and evaluation for glacier surface albedo retrieval from satellite observations. Remote Sens, 13(9). https://doi.org/10.3390/rs13091714

Reno MJ, Hansen CW, Stein JS (2012) Global horizontal irradiance clear sky models: implementation and analysis (Issue March)

Rn HK, Shafiullah GM, Maung A, Oo T, Stojcevski A (2014) Feasibility study of standalone hybrid power system modeled with photovoltaic modules and ethanol generator for victoria. Int J Eng Sci Invention 3(8):54–69

Resch B, Sagl G, Törnros T, Bachmaier A, Eggers J-B, Sebastian Herkel SN, H. G. (2014) GIS-based planning and modeling for renewable energy: challenges and future research avenues. ISPRS Int J Geo Inf 3(3):662–692. https://doi.org/10.3390/ijgi3020662

Rome CW, Shaddick G (2014) Sample size determination of photovoltaic by assessing regional variability Issue 0130, pp 11–12

Ruiz-Arias JA, Tovar-Pescador J, Pozo-Vázquez D, Alsamamra H (2009) A comparative analysis of DEM-based models to estimate the solar radiation in mountainous terrain. Int J Geogr Inf Sci 23(8):1049–1076. https://doi.org/10.1080/13658810802022806

Sadat SA, Faraji J, Babaei M, Ketabi A (2020) Techno-economic comparative study of hybrid microgrids in eight climate zones of Iran. Energy Sci Eng 8(9):3004–3026. https://doi.org/10.1002/ese3.720

Salisu S, Mustafa MW, Mohammed OO, Mustapha M, Jumani TA (2019) Techno-economic feasibility analysis of an offgrid hybrid energy system for rural electrification in Nigeria. Int J Renew Energy Res 9(1):261–270

Santos T, Gomes N, Freire S, Brito MC, Santos L, Tenedório JA (2014) Applications of solar mapping in the urban environment. Appl Geogr 51:48–57. https://doi.org/10.1016/j.apgeog.2014.03.008

Sarif MO, Gupta RD (2021) Spatiotemporal mapping of land use/land cover dynamics using remote sensing and GIS approach: a case study of Prayagraj City, India (1988–2018). Environ Dev Sustain. https://doi.org/10.1007/s10668-021-01475-0

Sawle Y, Gupta SC, Bohre AK (2016) PV-wind hybrid system: a review with case study. Cogent Eng 3(1):1–31. https://doi.org/10.1080/23311916.2016.1189305

Schwingshackl C, Petitta M, Wagner JE, Belluardo G, Moser D (2013) Wind effect on PV module temperature: analysis of different techniques for an accurate estimation. Energy Procedia 40:77–86. https://doi.org/10.1016/j.egypro.2013.08.010

Sevilla-Salcedo C, Gómez-Verdejo V, Olmos PM (2021) Sparse semi-supervised heterogeneous interbattery bayesian analysis. Pattern Recogn 120. https://doi.org/10.1016/j.patcog.2021.108141

Sharma P, Harinarayana T (2017) Solar energy generation potential along national highways. Int J Energy Environ Eng 4(16):1–13

Shi Z, Wang R, Zhang T (2015) Multi-objective optimal design of hybrid renewable energy systems using preference-inspired coevolutionary approach. Sol Energy 118:96–106. https://doi.org/10.1016/j.solener.2015.03.052

Shiva Chandra V, Reshma T (2022) Urbanization implications on local climate and groundwater levels using index-based techniques. Lect Notes Civ Eng 176:531–548. https://doi.org/10.1007/978-981-16-4629-4_36

Sinha AA, Shukla A, Prasad RB (2017) A review on CSP technologies with heat transfer fluids used in Indian power plants. In: International conference on 21st century energy needs—materials, systems and applications, ICTFCEN 2016. https://doi.org/10.1109/ICTFCEN.2016.8052711

Sinha G, Kronenfeld BJ, Brunskill JC (2015) Toward democratization of geographic information: GIS, remote sensing, and GNSS applications in everyday life. In Remotely sensed data characterization, classification, and accuracies vol 1. CRC Press. https://doi.org/10.1201/b19294

Tal Y, Hager BH (2015) An empirical study of the distribution of earthquakes with respect to rock type and depth. Geophys Res Lett 42(18):7406–7413. https://doi.org/10.1002/2015GL064934

Tanskanen A, Jantunen A, Saksa JM, Partanen J, Bergman J, Bertsch V, Treitz M, Geldermann J, Rentz O, Di Silvio B, Cesarotti V, Introna V, Babatunde OM, Babatunde DE, Denwigwe IH, Adedoja TB, Adedoja OS, Okharedia TE, Kwakwa PA, … Authors F (2021) New directions in UK energy policy. Int J Energy Sect Manage 15(2):157–172. https://doi.org/10.1108/IJESM-12-2019-0001

Tausendschön J, Radl S (2021) Deep neural network-based heat radiation modelling between particles and between walls and particles. Int J Heat Mass Transfer 177. https://doi.org/10.1016/j.ijheatmasstransfer.2021.121557

Thirumurthy N, Harrington L, Martin D, Thomas L (2012) Opportunities and challenges for solar minigrid development in rural India opportunities and challenges for solar minigrid development in rural India

Tijani HO, Wei Tan C, Bashir N (2014) Techno-economic analysis of hybrid photovoltaic/diesel/battery off-grid system in Northern Nigeria. J Renew Sustain Energy 6(3):033103. https://doi.org/10.1063/1.4873122

Tito SR, Lie TT, Anderson TN (2016) Optimal sizing of a wind-photovoltaic-battery hybrid renewable energy system considering socio-demographic factors. Sol Energy 136:525–532. https://doi.org/10.1016/j.solener.2016.07.036

Tudu B, Roy P, Kumar S, Pal D, Mandal KK, Chakraborty N (2012) Techno-economic feasibility analysis of hybrid renewable energy system using improved version of particle swarm optimization. In: Lecture notes in computer science (including subseries lecture notes in artificial intelligence and lecture notes in bioinformatics), vol 7677, LNCS, pp 116–123. https://doi.org/10.1007/978-3-642-35380-2_15

Utrillas P, Martínez-Lozano JA, Tena F, Pedrós R, Cañada J, Boscá JV, Lorente J (1998) Preliminary study of the influence of the urban effect on the spectral aerosol optical thickness in a Mediterranean coastal site. In: Proceedings of SPIE—The international society for optical engineering, vol 3495, pp 388–396. https://doi.org/10.1117/12.332694

Verma A, Biswas S, Ahmad SY (2015) Techno-Financial Analysis of Energy Access through Hybrid System with Solar PV under the Various Rural Community Models for State of Uttarakhand, India 75–94

Vinod Kumar TM, Krishnaveni KS (2020) Regeneration of smart spaces for smart living in Kozhikode Metropolitan Region (KMR). In: Advances in 21st century human settlements, pp 267–380. https://doi.org/10.1007/978-981-15-4615-0_5

Wafa ARA, Taha ATM (2014) Reliability evaluation of distribution systems under μ grid-tied and islanded μ grid modes using Monte Carlo Simulation 52–62

Wang Z, Cai J, Wang Q, Wu SC, Li J (2021) Unsupervised discovery of thin-film photovoltaic materials from unlabeled data. Npj Comput Mater 7(1). https://doi.org/10.1038/s41524-021-00596-4

Wilcox S, Gueymard CA (2010) Spatial and temporal variability of the solar resource in the United States. Am Sol Energy Soc Conf 1–8

Wu DD, Olson DL, Birge JR (2013) Risk management in cleaner production. J Clean Prod. https://doi.org/10.1016/j.jclepro.2013.02.014

Xiangdong L, Youlong H, Peng Q, Wei L, Bilalov RA, Smetannikov OY, Galanti L, Shasha D, Gunsalus KC, Rengaswamy N, Seshadreesan KP, Guha S, Pfister HD, Zhou L, Shen Y, Chen N, Li W, Lin HCHW, Zhou Y, … Miquel-Romero MJ (2021) J Bus Ethics 11(1):1–15. https://doi.org/10.1016/j.isprsjprs.2021.02.021

Xiao F, Shea GYK, Wong MS, Campbell J (2014) An automated and integrated framework for dust storm detection based on OGC Web processing services. In: Dragicevic SLS (ed) International archives of the photogrammetry, remote sensing and spatial information sciences—ISPRS Archives, vol 40, Issue 2. International Society for Photogrammetry and Remote Sensing, pp 151–156. https://doi.org/10.5194/isprsarchives-XL-2-151-2014

Xu S, Yang K, Xu Y, Zhu Y, Luo Y, Shang C, Zhang J, Zhang Y, Gao M, Wu C (2021) Urban land surface temperature monitoring and surface thermal runoff pollution evaluation using uav thermal remote sensing technology. Sustainability (Switzerland) 13(20). https://doi.org/10.3390/su132011203

Yang C-H, Lee K-C, Chen H-C (2016) Incorporating carbon footprint with activity-based costing constraints into sustainable public transport infrastructure project decisions. J Clean Prod 133:1154–1166. https://doi.org/10.1016/j.jclepro.2016.06.014

Yang X, Li Y, Luo Z, Chan, PW (2017) The urban cool island phenomenon in a high-rise high-density city and its mechanisms 37(2):890–904. https://doi.org/10.1002/joc.4747

Yariyan P, Avand M, Abbaspour RA, Torabi Haghighi A, Costache R, Ghorbanzadeh O, Janizadeh S, Blaschke T (2020) Flood susceptibility mapping using an improved analytic network process with statistical models. Geomat Nat Haz Risk 11(1):2282–2314. https://doi.org/10.1080/19475705.2020.1836036

Yi C, Kwon HG, Yang H (2022) Spatial temperature differences in local climate zones of Seoul metropolitan area during a heatwave 41. https://doi.org/10.1016/j.uclim.2021.101012

Zahid H, Umer F, Rashid Z, Raheem A, Shakoor R, Hussain GA (2020) Critical analysis and application of net-metering practice in MEPCO. Int J Photoenergy. https://doi.org/10.1155/2020/4135860

Zhang L, Na J, Zhu J, Shi Z, Zou C, Yang L (2021) Spatiotemporal causal convolutional network for forecasting hourly PM2.5 concentrations in Beijing, China. Comput Geosci 155. https://doi.org/10.1016/j.cageo.2021.104869

Zhang L, Xu Y, Yeh C-H, Liu Y, Zhou D (2016) City sustainability evaluation using multi-criteria decision making with objective weights of interdependent criteria. J Clean Prod 131:491–499. https://doi.org/10.1016/j.jclepro.2016.04.153

Zhao J, Yu L, Xu Y, Li X, Zhou Y, Peng D, Liu H, Huang X, Zhou Z, Wang D, Ren C, Gong P (2020) Exploring difference in land surface temperature between the city centres and urban expansion areas of China's major cities. Int J Remote Sens 41(23):8963–8983. https://doi.org/10.1080/01431161.2020.1797216

Zhao TF, Fong KF (2017) Characterization of different heat mitigation strategies in landscape to fight against heat island and improve thermal comfort in hot–humid climate (Part I): Measurement and modelling. Sustain Cities Soc 32:523–531. https://doi.org/10.1016/j.scs.2017.03.025

Zhou Y, Elchalakani M, Liu H, Briseghella B, Sun C (2022) Photocatalytic concrete for degrading organic dyes in water. Environ Sci Pollut Res. https://doi.org/10.1007/s11356-021-18332-2

Chapter 4
Conventional Methods of Energy System Evaluation

Abstract The data on solar radiation may be used to calculate the amount of energy that is delivered by the sun to the surface of the planet at a certain place and during a particular time. These kinds of data are essential for effectively researching solar energy. Because of the high cost and limited capacity of physically in-situ data harvesting of solar radiation, we are unable to get a regular dataset that is also made accessible to the public. As a result, we need to develop alternative ways to acquire this data. The current part of the investigation involves doing an in-depth analysis of solar energy modelling approaches, which are arranged following the kind of information they provide. In addition to this, a variety of models and methods that were created or sponsored by a variety of organizations to evaluate, analyse, and optimize energy efficiency and renewable energy systems have also been included. Some of these tools may be used on a worldwide scale, on a regional scale, on a local scale, or for a particular purpose. These approaches provide a scientific solution to the issue at hand while also shedding light on how scientific research is carried out as a discipline. In addition to that, a variety of different data processing methods were developed and put into practice to deal with the raw data or intermediate data that were produced during the process of modelling. Following that, all the data for the region's sample solar maps were integrated and plotted or otherwise created. In a word, what we did was study the many methods that a researcher may use to investigate this kind of research challenge, in addition to the reasoning behind those methods. The researcher must be familiar with not only the study methodologies and procedures, but also the ideas since similar aspects of different problems might provide unique challenges.

Keywords *Energy modelling* · *Data processing* · *Meteorological and statistical* · *SolMap*

4.1 Overview

Solar energy is clean and renewable. It may supply heat, light, power, and solar heating, and cooling. The sun's power to create renewable energy includes producing wind for wind power and growing plants for biofuels. Solar energy is simple to use

D. Kumar, *Renewable Energy Scenarios in Future Indian Smart Cities*, Advances in Geographical and Environmental Sciences, https://doi.org/10.1007/978-981-19-8456-3_4

at home, which is a plus. You do not need a solar farm or a lot of money to profit from solar electricity. Solar panels make solar electricity easy and economical no matter where you live or your budget. Solar panels can help clean up metropolitan sky. They emit no hazardous pollutants and minimize building energy use (Santos et al. 2014). Most solar panels developed today can withstand severe winds and heavy snow, making them useful in practically any environment. It is essential to conduct measurements and create models of both broadband and spectrum terrestrial solar radiation before beginning the process of evaluating and installing solar renewable energy systems. The term "solar energy" refers to the process of transforming the radiant energy of the sun (both directly from the solar disc and indirectly from the sky) into a variety of types of energy that are usable by humans. Incoming sun insolation is essential to most physical and biological processes because of the role it plays in maintaining (Khandelwal et al. 2016; Khorrami et al. 2021). People are exploring alternative energy sources since the supply of non-renewable resources such as fossil fuels and other non-renewable resources are decreasing. Solar energy looks to be a potential solution to this issue since it is a renewable energy source that produces less pollution and is also capable of being quickly converted into electricity through photovoltaic (PV) systems.

In the current study, an explanation of how to estimate solar radiation using multiple distinct models is provided. These models include the Direct Isolation Simulation Code (DISC), the METSTAT (Meteorological and Statistical), and the Climatological Solar Radiation (CSR) models for estimating solar radiation from meteorological data (Kumar 2020; Tanskanen et al. 2021). Other models include the Bird clear sky model for global, direct beam, and diffuse terrestrial solar radiation. The amount of solar radiation is estimated using each of these models. Broadband solar radiation model techniques have been developed to predict direct beam, total hemispheric, and diffuse sky radiation (Chwieduk 2009; Gambhir et al. 2010). One such model approach was produced at the National Renewable Energy Laboratory (Myers 2005). Processing the values received from the spatial radiation models made it possible to create more granular datasets. The datasets were validated, sorted, summarized, aggregated, analysed, reported, and classified by making use of a wide variety of data processing procedures, which ranged from manual to automated to electronic methods. These methods included: aggregating, summarizing, analysing, reporting, and verifying the data. This was done so that the great data would have all the desirable attributes (Agugiaro et al. 2005; Liu 2009; Meisen et al. 2006).

We concluded that creating a cartographic depiction of solar insolation parameters would be the best way to get a deeper comprehension of the global distribution of solar potential, PV output, air temperature, and other spatial aspects (Bechtel et al. 2015; Chrysoulakis 2002; Sumari et al. 2020). This will be of help to us. There was also a yearly map and a map that presented the locational data of the measurement sites, in addition to the monthly maps and the map that displayed the data. The sun's rays that reach the surface of the earth are the product of a series of complicated energy transfers that take place between the atmosphere and the surface of the planet (Gómez-Chova et al. 2017; Goodchild 2009). Radiation gradients at different latitudes are produced on a global scale because of the shape of the planet, as well as its rotation and revolution around the sun (Narimani et al. 2021; Padmanaban et al. 2019).

The major elements that determine the dispersion of radiation at regional and local scales include variations in elevation, surface inclination (slope), and direction (aspect), as well as shadows created by terrain features (Li et al. 2020; Velotto et al. 2016; Wang et al. 2021). The spatial and temporal variation of the amount of solar energy that is received plays a significant role in determining the dynamics of a wide variety of landscape processes. These processes include the temperature and moisture content of the air and soil, the melting of snow, photosynthesis, and evapotranspiration (Abujayyab et al. 2021; Anderson et al. 2008; Mhawej et al. 2021). This variation also has a direct impact on human society. Sun radiation data that are precise and have a good geographical distribution are needed for a variety of applications, including environmental research, climatology, ecology, architectural design, remote sensing, photovoltaics, and land management, amongst others (Hwang and Liu 2008; Loukanov et al. 2020; Peng et al. 2018; Rao et al. 2005).

4.2 Sub-objectives

1. Modelling of solar radiation through spatial solar radiation (S^2R) model.
2. Processing of solar radiation data assimilated from different sources.
3. Solar insolation map generation to visualize solar energy potential.

4.3 Workflow Overview

The data that we collect on solar radiation can tell us how much of the sun's energy strikes a surface at a particular place on the globe at a certain point in time. This knowledge is essential for doing fruitful research on solar energy (Ajayi et al. 2016; Rn et al. 2014; Sawle et al. 2016). The high cost of getting in-situ data, the difficulty of measuring solar radiation, and the lack of data that are already prepared to stimulate the interest of researchers in developing novel ways for obtaining this data. Within this component of the research, there is also a comprehensive investigation of the various modelling approaches for solar energy (Dawson et al. 2020; Delkash et al. 2018; Luo and Paal 2021). For the most part, the study focused on taking into consideration linear, nonlinear, and artificial intelligence models for estimating solar energy (Khatib et al. 2012; Weng and Hu 2008).

In Fig. 4.1, the general workflow of the methodology used in this chapter is depicted. To make the work easier to accomplish, the work was divided into three parts. The core subject of the study, as shown in Fig. 4.1's workflow, is the modelling of solar resources using the defined approaches, which can be further expanded to

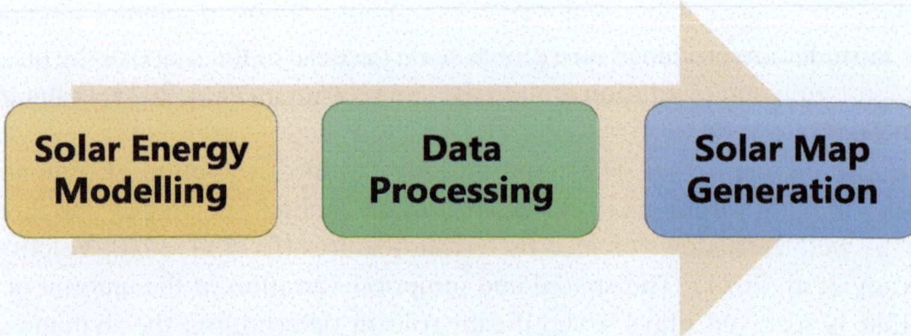

Fig. 4.1 Schematic for workflow diagram

include data processing and solar map production (Kumar et al. 2008; Meisen et al. 2006). The study's first section dealt with the selection of the study region and the collecting of necessary data, which was followed by data modification to ensure compliance with the chosen solar software for analysis. To mimic the incident solar radiation, the altered data were then input into the chosen solar software. After that, the simulated data were processed using data processing techniques to create the essential datasets for solar map production (Fadare 2009; Kumar 2020).

4.4 Methods

This segment discusses the methodology, data sources, and software/tools that are used to complete the task at hand. The present phase of the study makes use of both quantitative and analytical research approaches to gather information. Analytical and quantitative research approaches both involve aspects of the estimate and selection of processes utilized in data analysis (Tajbakhsh and Shamsi 2019; Yang et al. 2020). The analytical design is an essential component of the study design since it enables us to determine the method by which the data will be assessed, in addition to providing information on the repercussions of the data or observation and the ultimate result. Quantitative research techniques rely on a wide range of mathematical and statistical models and algorithms to arrive at the desired conclusions. Analysing the situation and gaining an understanding of the connection between information that has been visually interpreted or digitally processed and the real situation or phenomena that is taking place on the ground is another kind of analytical technique (Arefin 2020; Yang et al. 2019).

Quantitative methods are also essential for determining the values of a certain parameter. In any modelling approach, the first step was to determine the factors that would help determine the value of the thing being estimated (Ahmad and Thaheem 2018; Rao et al. 2021). This section also highlighted the importance of the percentage of sunshine, cloud cover, and air mass when it comes to the modelling of diffuse irradiance. Its worldwide analogue, which has been developed in the past and is recognized in this article, is the most essential statistic for measuring diffuse irradiance. As stated in earlier section, this relationship is not generic, and to fully understand it, models, that consider the geographical and climatic conditions of the area in question are necessary (Chu and Guo 2013; Vinod et al. 2018). In addition to the global radiation, it was found that other meteorological components also had a significant influence on the estimate of the diffuse irradiance, which led to the improvement of the accuracy of the calculation (Barron et al. 2008; Kumar et al. 2017).

The modelling of incident solar energy under circumstances of complicated sky occlusion is performed in the same way for each location on a topographic surface. The irradiance that is emitted from clear sky directions is assessed in terms of its direct and diffuse components after being integrated across one hemisphere's worth of sky directions (Zhao et al. 2008). However, there are still a lot of obstacles to overcome, both in terms of modelling and in terms of measurement. As a result of the models' inability to consider the significant temporal variability of the sky conditions, the anisotropic irradiance distributions, the geometric complexity of plant canopies, and the intricate patterns of reflection (scattering) that result from this interaction between the various canopies and their surroundings.

The main selling feature of the approaches was the actual daily incident solar insolation shown as a function of latitude, year, and likely daylight hours. An ArcGIS-based solar energy model was used to conduct the primary study, which consisted of analysing and visualizing the influence that the sun has on a geographic region throughout a certain amount of time. The models of insolation were developed taking into consideration factors such as air conditions, elevation, surface orientation, and topographic influences. The primary emphasis of our research was on developing an application for this model of insolation that would allow for the spatial interpolation of intricate topography at landscape sizes. The existing techniques for interpolation are generally utilized in continental or large regional regions; however, they do not capture the enormous diversity that occurs at smaller scales. Measurements taken on the ground are more reliable than those extrapolated from satellites, thus it is best to utilize them wherever possible. These tactics, on the other hand, are extremely helpful despite the limited availability of these data (Kumar et al. 2008; Sawle et al. 2016).

The two forms of observations that may be used to collect data on solar radiation are satellite and ground-based observations. Satellite-based direct normal irradiance (DNI) or data suppliers include organizations such as NASA, Meteonorm, NREL, 3Tier, ISHRAE, and Weather Analytics. Although data for the entire country have not yet been recorded, the in-situ data that have been documented by organizations in India such as the India Meteorological Department (IMD) stations and the National Institute of Wind Energy (NIWE) stations are included in the ground-based radiation data. These organizations include the India Meteorological Department (IMD) stations.

4.4.1 Location of Research Area

The experimental research was performed for the Kalaburagi district, which was previously well known as the Gulbarga district. This district is positioned in the northern division of the state of Karnataka. Its coordinates are as follows: 76°04′ to 77°42′ East Longitude and 17°12′ to 17°46′ North Latitude. Its total land cover area is 10,951 km^2. The normal elevation of the area is somewhere between 300 and 750 m above mean sea level. Both the Krishna and the Bhima rivers are quite significant in this area (Kumar 2019). The climate in this area has three distinctly different seasons each year. Temperatures throughout the summer range from 38 to 44 degrees Celsius and begin in late February and continue until the middle of June. The southwest monsoon follows, lasting from the latter half of June to the latter half of September, and bringing with it substantial rainfall of up to 750 mm (27 to 37 °C). It is then followed by dry winter weather with temperatures ranging from 11 to 26 C until the middle of January. Figure 4.2 is an illustration of the geographical setting of the study region.

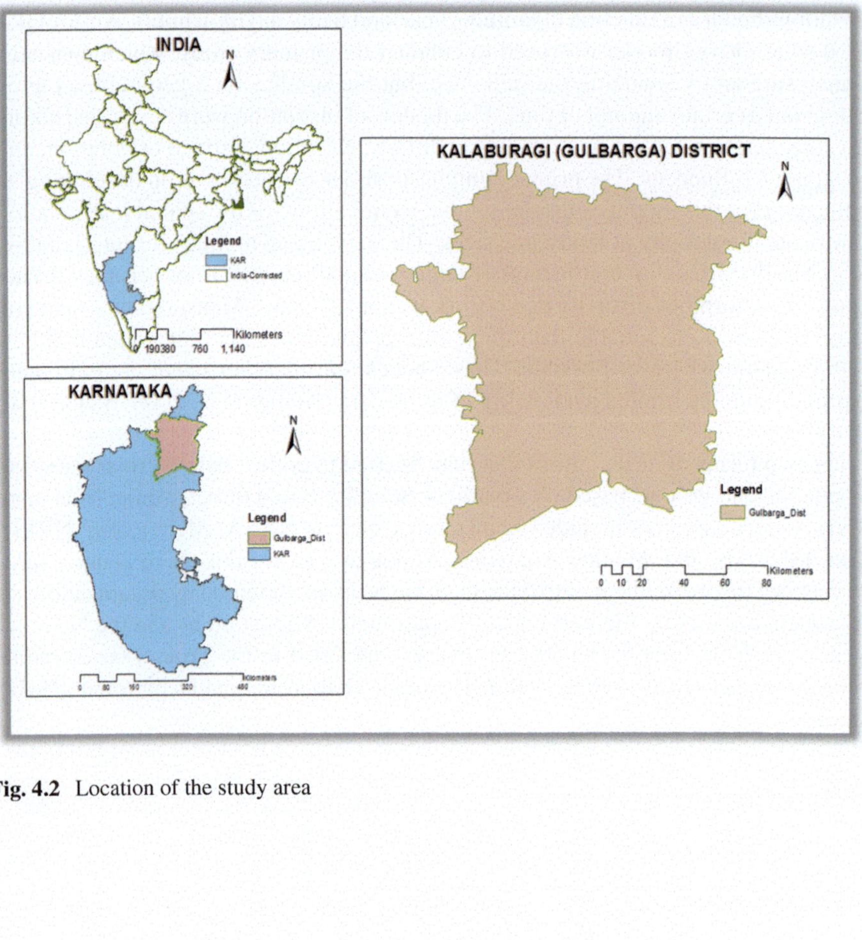

Fig. 4.2 Location of the study area

Why this research location?
To broaden the scope of the study being done, a section of an experimental district was selected. Former chapter presents a brief explanation of the factors that led to the selection of Kalaburagi as the site for the experimental research. A few of the more general explanations are as follows:

1. An absence of fuzzy borders is a self-contained zone.
2. A diverse and highly unstable characteristic of the terrain, with a variety of biological and landscape zones.
3. The author is quite familiar with the region as well.

4.4.2 Datasets Used

In this study, we employed GCPs that were obtained using a handheld GPS, a topographic sheet from the Survey of India with a scale of 1:50 K, SRTM DEM/BHUVAN DEM, and very high-resolution (VHR) satellite data from Google Earth (site validations). Additional open archive data sets for execution and validation are available from BHUVAN NOEDA (NRSC), Landsat Archive, and Earth Explorer (USGS). The information was gathered from multiple open source or educational Websites along with in-situ data. The ancillary facts were accumulated from books, journals, reports, and Websites. The various information sources extensively referred to include:

1. Online gateways and databases *(ResearchGate).*
2. Ministry of New and Renewable Energy *(Website, Reports, Manuals).*
3. Online Journals *(e.g., Solar Energy, Energy, Journal of Solar Energy Engineering).*
4. Books in the context of infrastructure management, trends in energy sciences.
5. Reference texts from ESRI, **ERDAS Inc.** publications and manuals.

4.4.3 Software and Tools Used

1. Energy Plus 8.0
2. Geospatial Toolkit *(NREL)*
3. RET Screen Plus Software
4. ESRI ArcGIS 10. x (*for vector data analysis*)
5. ERDAS 201x (*for image processing*)

4.4.4 Modelling of Solar Radiation

Applications based on solar energy need available solar data that are both location-specific and long-term. However, despite the significance of measured diffuse irradiation, it is usually impossible to get it; hence, several regression techniques have been developed to predict it using satellite data that are more readily accessible (Alrwashdeh 2018; Borfecchia et al. 2014). In addition to global for diffuse irradiance modelling, the research methodology being developed in this section includes the inclusion of constraint parameters such as latitude, day of the year (DOY), local terrain configurations, and atmospheric and climatological parameters such as sunshine fraction (SF), cloud cover (CC), and relative air mass (m).

The results of this study are presented and discussed in great depth utilizing datasets obtained from the experimental locations is already described. The latter offers a comprehensive summary of the method that was carried out to guarantee that the data that are being looked at are of good quality. The subsequent section covers up the issues of daily data and modelling, whereas this chapter concentrates on the hourly or sub-hourly values of horizontal radiation. Hourly radiation is needed for realistic models of solar energy systems, which need hour-by-hour computations, rather than monthly averages or daily readings. This makes hourly radiation an essential variable. In addition, radiation measurements collected on an hourly basis contribute to a more accurate assessment of the operation of a solar power system. As a result of the interplay of four factors inside the models, each combination that could be thought of was employed to develop its own set of models. In the end, the model that was deemed to be the most effective for the study locations was selected. After considering both the gain inaccuracy and the optimal number of input variables, we came to this conclusion.

The application will determine how often it is necessary to gather data on the amount of radiation received from the sun. However, for intensive modelling studies, hourly or sub-hourly estimations of inclined surface irradiation are necessary. For agricultural meteorology, a monthly or even an annual energy budget would be adequate. On the other hand, daily data are very necessary for the day-to-day monitoring of performance in applications such as solar gains from vertical glass, daylighting, and agricultural activities. In addition to daily radiation measurements, hourly or sub-hourly radiation data are routinely condensed into daily forms for usage in a variety of different applications that are only briefly concerned with solar energy. These daily forms are used for usage in a variety of different applications that are used in conjunction with daily radiation measurements.

Modelling is a fundamental and quantitative method for comprehending and analysing complicated systems and phenomena, as we all know. When trials are: (i) too big, (ii) too expensive, (iii) too risky, or (iv) too time-consuming, it is frequently utilized in place of them. It is becoming increasingly prevalent in the domain of computational, earth and space system modelling due to developments in various modelling tools and approaches.

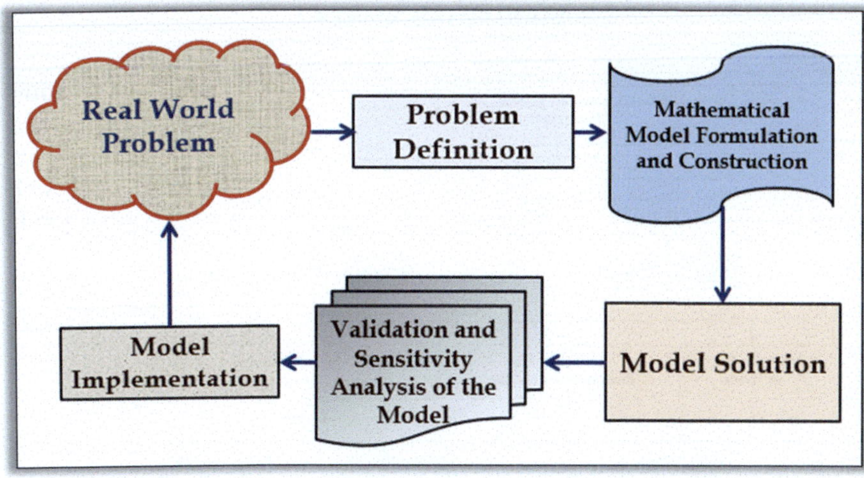

Fig. 4.3 Modelling pipeline *(Conceptual Map for model development concept)*

These developments have brought more clarity to the comprehension of any of the modelling schemes, which has contributed to the field's growth in popularity. The following diagram, Fig. 4.3, illustrates the fundamental structure of a modelling pipeline. It provides a visual representation of the many steps that are required for every modelling system.

This study is a step in the right direction since it underlines the difficulties associated with model validation in addition to offering spatial solar radiation models that take into account sunshine duration, cloud cover, and global radiation. It was determined that the dataset might significantly benefit from the addition of factors such as SF and/or CF, which led to the discovery of the significant improvement. However, because there are a greater number of measurable parameters, it is challenging to generalize the model. This is because there will be a greater number of measurement imperfections, which will lead to an increase in performance ambiguity, especially after applied to an independent dataset. Figure 4.4 is a diagram that illustrates the many types of solar radiation models.

(a) *Actual Incident Solar Radiation Model*
The variations in space and time of the actual incident solar radiation are primarily determined by several variables, including the following: (i) latitude, (ii) day of the year (DOY), (iii) configuration of local terrain, and (iv) atmospheric transmittance, and it has been also studied in (Hassan et al. 2011). For computation, the daily solar radiation on a horizontal plane at a point of interest is solved using the following equations, as recommended by (Hassan et al. 2011).

$$R_{\text{daily}} = (a + b\frac{n}{N}) * R_{\text{daily - ex}} \tag{4.1}$$

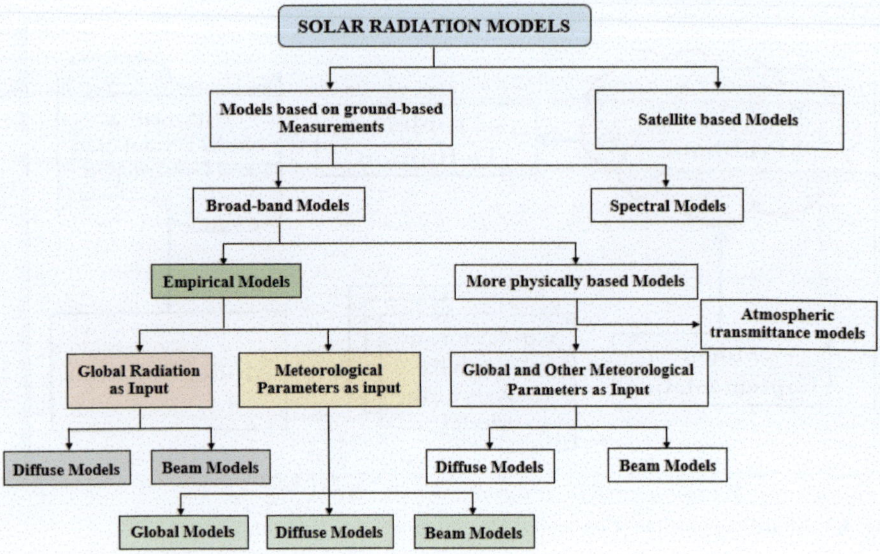

Fig. 4.4 Network diagram for illustrating solar models

$$R_{\text{daily - ex}} = \frac{1}{\pi} E_o d_r [\omega_s \sin(\phi) \sin(\delta) + \cos(\phi) \cos(\delta) \sin(\omega_s)] \tag{4.2}$$

$$d_r = 1 + 0.33 \cos(\tfrac{2\pi}{365} DOY) \tag{4.3}$$

$$\delta = 0.409 \sin(\tfrac{2\pi}{365} \text{DOY} - 1.39) \tag{4.4}$$

$$\text{Radians} = \tfrac{\pi}{180} [\text{decimal degrees}] \tag{4.5}$$

$$\omega_s = \arccos[-\tan(\phi) \tan(\delta)] \tag{4.6}$$

where R_{daily} and $R_{\text{daily-ex}}$ are the daily and extra-terrestrial incident radiation, respectively [$W \cdot m^{-2}$], a (=0.25) and b (=0.50) are the Angstrom values [dimensionless], n/N is the % of possible daylight hours (i.e., form of bright sunshine hours) [dimensionless], E_0 is the solar constant = 1353 $W \cdot m^{-2}$, d_r is the inverse relative distance between earth-sun, δ is the solar declination angle [radians], φ is the latitude [radians], ω_s is the sunset hour angle [radians], and DOY is the number of the day in the year between 1 (1 January) and 365 or 366 (31 December). Equation (4.1) can be used to calculate the daily potential incident solar radiation ($R_{\text{daily-potential}}$, $W \cdot m^{-2}$) if we consider the value of n/N (=1) and take the following form:

$$R_{\text{daily potential}} = 0.75 * R_{\text{daily - ex}} \tag{4.7}$$

(b) *Spatial Solar Radiation (S^2R) Model*
Incoming solar radiation, also known as insolation, passes through the atmosphere and is influenced or altered by the topography and surface characteristics of the planet before being intercepted as direct, diffuse, or reflected components at the earth's surface. Radiation that comes directly from the sun may travel unimpeded great distances before being stopped. Clouds and dust are examples of atmospheric components that may deflect diffuse radiation. The properties of the surface reflect the radiation that has been reflected. The sum of the sun's direct, diffuse, and reflected rays is referred to as global or total solar radiation. Figure 4.5 depicts the input window of the model that was constructed, which would calculate solar radiation depending on a variety of inputs (Ostapowicz 2009).

Several presumptions must be made when modelling solar energy over occupied regions, including (i) the estimation of existing or prospective solar radiation, (ii) the technological means of generating energy from incoming solar radiation, and

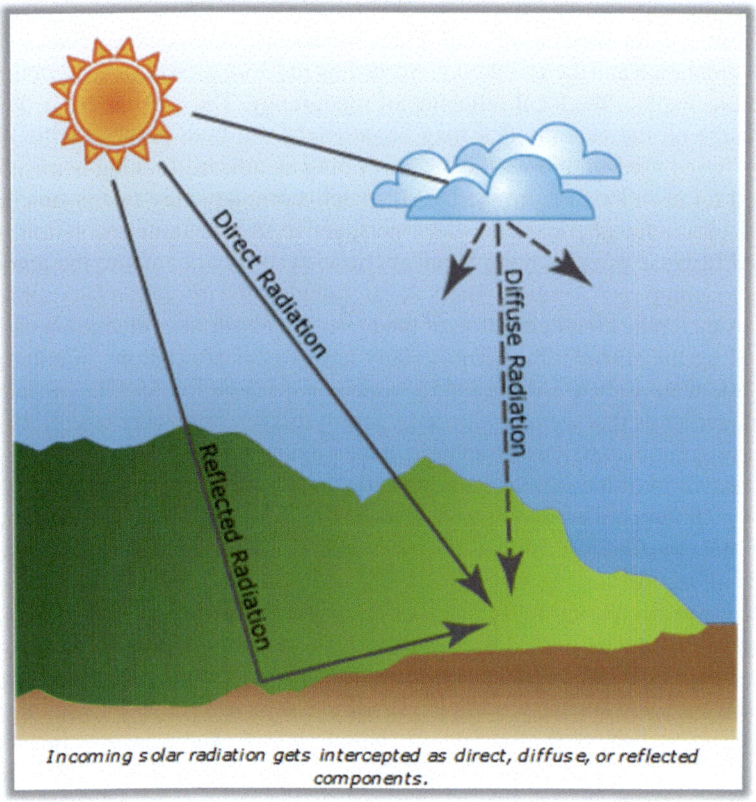

Fig. 4.5 Illustration showing solar radiation basics. *Source ArcGIS User Manual*

(iii) the suitable surface area inside a built-up region that is most exposed to the sun. As a result of the high expense of conducting radiation measurements, it is obvious that not only is it necessary to develop models of solar radiation for use at a variety of timestamps but there should also be continual efforts to enhance the accuracy of predictions. Both models for the global distribution of solar irradiation in terms of area and location were developed. It is not feasible to use the same models to estimate daily radiation levels since the models generated from hourly based data are different from those acquired from hourly based data.

To phrase it another way, models are strictly spatio-temporal and cannot be switched out for one another. Because of this, models for estimating radiation are created throughout this chapter, and an effort is made to assess these models using daily data. The quantity of solar energy that occurs at a certain location on the surface is mostly governed by the topography at that location. Large local solar radiation gradients may be produced when elevation, slope, slope direction (aspect), and shadowing all vary. These radiation gradients have the potential to influence a wide range of biophysical processes directly or indirectly (Dubayah and Rich 1996). There are two distinct varieties of spatial models for solar radiation and insolation: point-specific and area-based models. When calculating insolation for a particular place, point-specific models are used. These models are based on the geometry of the surface orientation and the visible sky. According to (Bornatico et al. 2013), empirical connections explain the local influence of topography. The S^2R model is designed to capitalize on the benefits that may be derived from both point-specific models and area-based models. It does this by generating an upward-looking hemispherical viewshed for each location on the DEM, so delivering a picture that is functionally identical to a hemispherical or fisheye perspective. An accurate insolation map is produced because of employing hemispheric viewsheds to compute the amount of sunlight reaching each location. The S^2R can calculate the insolation over some time, as well as the latitude and elevation of the location, the surface orientation, shadows produced by the surrounding terrain, daily and seasonal variations in solar angle, and air attenuation. It was designed to work as an add-on for ArcGIS. Because the software and tools that were produced as part of this research were modifications of original programmes written elsewhere, they are not included in this work.

Because most of the software and tools that were produced as part of this project were only modifications of preexisting programmes, they are not included in the volume that constitutes the thesis. On the other hand, adequate references have been included alongside any use of original programmes when it was deemed necessary to do so. It was tried to apply the real incidence solar radiation model; however, there were not enough meteorological data sets to make it possible (especially solar radiation data from IMD). As a direct consequence of this, solar insolation was simulated with the use of geospatial technology. Empirical computation is the approach that is utilized to fuel the S^2R model that is based on GIS. The estimates included, in addition to empirical techniques, (i) analytical methods, (ii) satellite estimations, and (iii) stochastic methods (kriging). Using empirical methods to model solar radiation required performing operations such as the following:

The different methods involved in the design of any solar radiation model are detailed in Fig. 4.6. The design and building of the required database are the first step in the process. In our scenario, the file geodatabase was created to store the geospatial data type in the database. Following that, the toolbox is constructed, followed by the toolset, and lastly, the matching model is prepared by combining or consolidating all these elements. This framework is supplemented with two extra checkpoints, namely edit the model and run the model, to make it easier for the user to deal with the model. The core of the entire process, as shown in Fig. 4.6, is the creation of a toolbox, toolset, and model. Figure 4.7 shows the actual model that was formed once the operations were completed. The presence of components in the model is demonstrated in this diagram. The first component (shown by the blue hue) is input, which receives human input. This is linked to the processing component (in yellow) that performs the desired operation on the input. The system is currently attempting to compute area-wise solar radiation values. The output units, which are likewise connected to the processing unit of the specified model, are the final component. It allows the viewer to see the many output types, such as direct radiation raster, diffuse radiation raster, and direct duration raster, as well as a composite image of region-wise solar radiation. The ArcGIS model builder was used to create this model.

After saving and attaching the model to one of the existing toolsets in the toolbox, you must complete the required steps after building the model in ArcGIS model builder. It is now ready to be put into action. Once the model has been run (which can be done by just clicking on the toolset), it will prompt the user for input in the form of a raster image (which is a high-resolution DEM image for better results) and a set of parameters to be set for computing the desired results. Figure 4.8 depicts the model's input window. We can see in the diagram that it is asking for several parameters as well as possibilities to create topographic and temporal limitations.

4.4.5 *Statistical Indicators for Evaluation of Model*

The following statistical measures were used in the evaluation of the efficacy of models for specific regions about datasets made accessible from a variety of sources:

a. Coefficient of determination (R^2)
b. Mean bias deviation (MBD)
c. Root means square deviation (RMSD)
d. Skewness
e. Kurtosis.

The coefficient of determination (R^2) is the proportion of variance that can be explained as a total variation. It is somewhere between 0 and 1. It is preferable to have a high R^2 level since it shows less variation that cannot be explained. R^2 is a statistic that is widely used to evaluate the appropriateness of a regression model; nevertheless, it should not be utilized as the main basis for choosing a model.

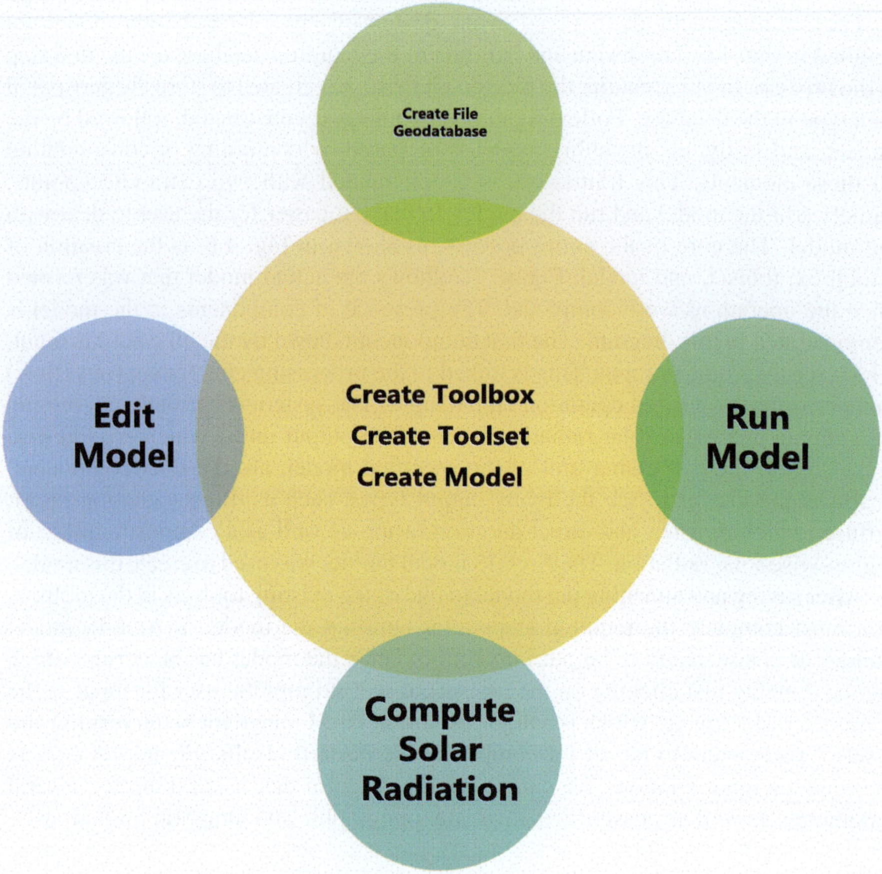

Fig. 4.6 Illustration for solar radiation model design

Mean bias deviation (MBD) gives an indicator of the model's overall trend over a longer period. When the MBD number is positive, it indicates that the suggested model overestimates the data, whereas when it is negative, it indicates that the model underestimates the data. The MBD ought to be about 0 if at all feasible. (MBD) is a statistic that is used to measure how accurate the model is overall. It determines how likely the model is to either underestimate or overestimate the values that were measured, and it must be reset to zero. The following contribute to its value:

$$MBD = \frac{1}{N}\sum_{i=1}^{n}(X_p - X_m) \tag{4.8}$$

where X_p is the '**i**'th predicted value and X_m is the '**i**'th measured value.

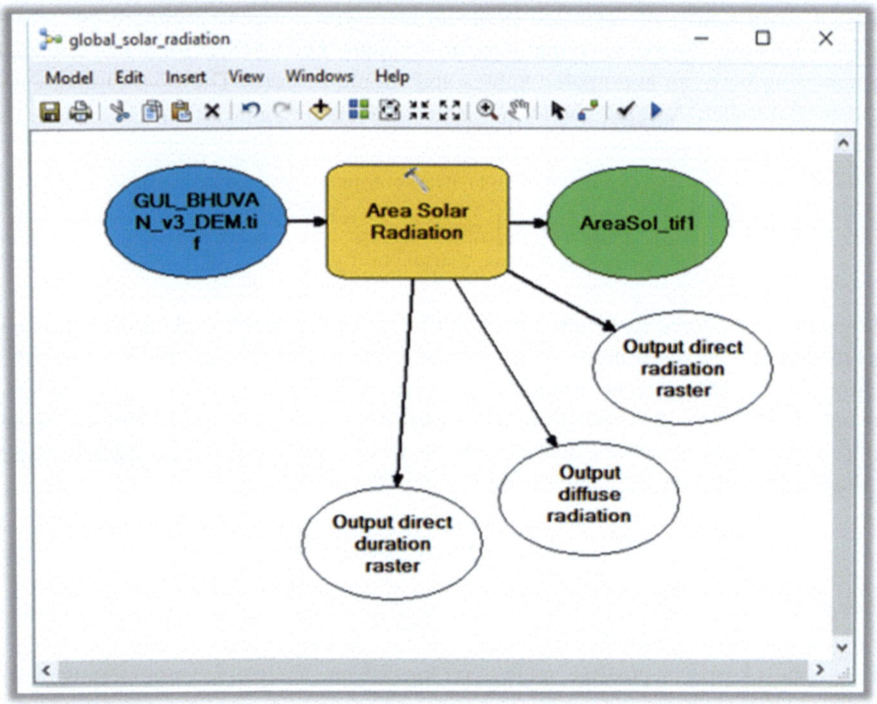

Fig. 4.7 Illustration showing developed solar radiation model

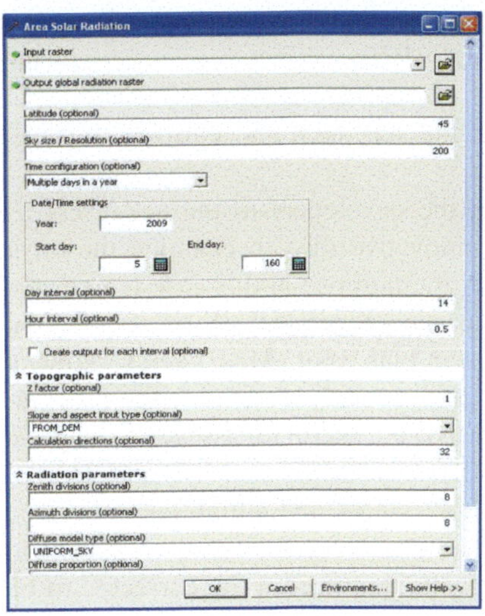

Input: DEM (X m)

Output: radiation raster's (watt hours per square meter), duration (hours).

Diffusion proportion: the proportion of global normal radiation flux that diffuse (0.3 generally clear sky conditions.

Transmissivity: the fraction of radiation that passes through the atmosphere

0 – no transmission,

1 – transmission.

0.5 – generally clear sky.

Fig. 4.8 Illustration of the model window for seeding input parameter

Root mean square of deviation (RMSD) gives you a way to compare the actual variance between the expected and measured values. The RMSD calculates the level of scattering between anticipated values and should be zero. The RMSD is always a positive number. A better model has a lower absolute value of RMSD. It can be expressed as:

$$RMSD = \left[\frac{\sum (X_p - X_m)^2}{N}\right]^{1/2} \tag{4.9}$$

Skewness: The degree to which a distribution lacks symmetry is determined using this measure. If the distribution seems the same to the left as it does to the right of the centre point, it is said to be symmetric or normal. When the symmetry is finished, the outcome is zero. A distribution that is positively skewed would have a tail at the high end of the scale, while a distribution that is negatively skewed would have a tail at the low end of the scale. The expression for skewness is as,

$$\text{Skewness} = (\text{mean - mode})/\text{standard deviation} \tag{4.10}$$

A distribution's kurtosis is a measurement of how favourably skewed the distribution is relative to its width. For a normal distribution, the kurtosis statistic has a value of zero (mesokurtic), a positive value (leptokurtic), and a negative value (leptokurtic) for flat distributions (platykurtic). The following equation provides a quartile- and percentile-based numerical representation of kurtosis,

where $Q = (Q_3\text{-}Q_1)/2$ is the semi-inter quartile range.

$$\text{Kurtosis} = Q/(P90 - P10) \tag{4.11}$$

When put into practice, the true values of the skewness and kurtosis statistics rarely turn out to be precisely zero.

Everyone engaged with the models, from the developers to the end users, is concerned about estimation errors. It is an innovative idea to examine the error statistics related to the radiation calculation as the purpose in this case is to arrive at a radiation estimate. The most often used measures for statistically evaluating the efficiency of solar radiation models are the mean bias error (MBE) and root mean square error (RMSD). Skewness and kurtosis, respectively, determine the distribution's symmetry and shape, while MBD and RMSD are useful for assessing radiation forecast performance. The abbreviations for MBD and RMSD are MBD and RMSD, respectively. It is ideal to have a low absolute skewness and a high kurtosis since the study takes the cause of error into consideration. Nevertheless, each statistical indicator has its own unique set of inherent faults that need to be corrected to be considered reliable. The following is a list of some of them:

(a) The underestimate of one observation in a data collection balances out the overestimation of another observation in the same data set. Because of this, MBD struggles to accurately depict the true nature of model accuracy.

(b) The RMSD does not provide a thorough assessment of a model since only a few significant errors might significantly increase the model's worth.
(c) Adding an extra parameter or higher-order polynomial components to the regression equation might artificially boost the coefficient of determination (R^2), which is not necessarily a good indicator of how applicable the model is. Furthermore, the presence of a high R^2 does not guarantee that the regression model will be able to predict future data accurately (Yang et al. 2021).
(d) Skewness just provides information about the normality of the distribution; it does not account for the extent of the mistakes.
(e) If measured values are included in the data set, the existence of outliers may make the kurtosis statistic very sensitive. In other words, a few errors in the distribution's tails, which cause a divergence from false or irrelevant data, might significantly affect the distribution's value (Wahlström et al. 2008).

As a result of these shortcomings, there will always be a disparity when assessing the performance of a model until a technique of assessment that is consistent is developed. The statistical analysis will, in future, serve as the basis for the proposal of indigenous indices such as precision score. This analysis will also include conventional indicators and indices. A model that has not been confirmed before being submitted is worthless (Table 4.1).

When applied to a new collection of data using the same statistical methods, a model that is based on the statistical analysis that was selected could not give encouraging conclusions. Therefore, rather than proposing models based only on how well they work when applied to the same data, it is advisable to put into practice the evaluation of the models by applying them to a data set that is entirely different from the original data.

where Site 1 is Object ID2, Site 2 is Object ID6, Site 3 is Object ID7, Site 4 is Object ID8, Site 5 is Object ID10, and Site 6 is Object ID18 respectively. All other values of NREL and TAMU models are within a major grid covering all the points inside, as shown in Fig. 4.9.

The research and development of solar energy conservation systems cannot proceed without the collecting of data on solar radiation. To arrive at an estimate

Table 4.1 Comparative solar radiation values (annual) from different models

Location	Yearly (kWh/m^2/day)				Average bias *RET screen Vs Model*	Precision score (PS)
	Model	NREL	TAMU	RET screen		
Site 1	7.439236	5.62618	5.3793765	6.828756784	0.610479	−0.59087961
Site 2	7.537281	5.62618	5.3793765	6.828756784	0.708524	−0.59087961
Site 3	4.067364	5.62618	5.3793765	6.828756784	−2.76139	−0.59087961
Site 4	7.280376	5.62618	5.3793765	6.828756784	0.451619	−0.59087961
Site 5	5.226242	5.62618	5.3793765	6.828756784	−1.60251	−0.59087961
Site 6	5.876764	5.62618	5.3793765	6.828756784	−0.95199	−0.59087961

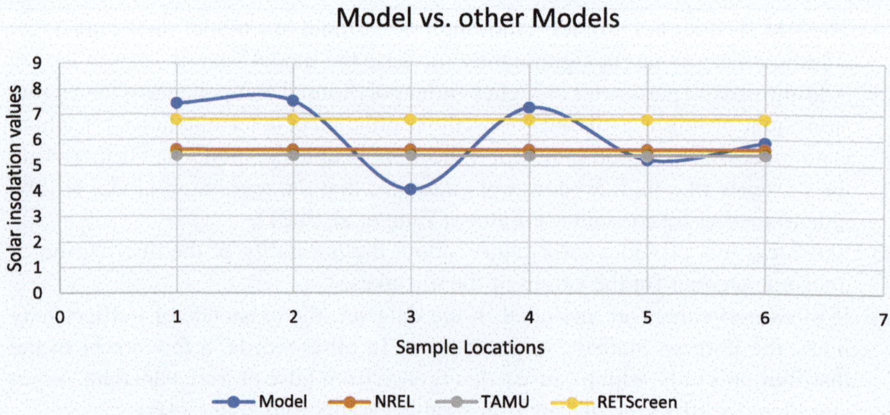

Fig. 4.9 Plot illustrating model versus other models' data results

of the monthly average daily global radiation on a horizontal surface, empirical correlations need to be constructed first. Models that consider the amount of sunlight in a region are used in the process of estimating total global solar radiation for a location. Researchers interested in solar energy will be able to put their faith in the estimated data thanks to the correlation equations that were described in this study work. The calculated data and the observed data correspond quite well. Some of the possible findings and inferences that might be drawn from this study are as follows:

1. Solar energy technologies are crucial components of a sustainable energy future since they provide a source of energy that is not only clean, but also renewable, and they can be utilized inside the confines of the house. In order to design and evaluate solar energy systems, one requires information on the components of solar radiation at a particular place. This information may be obtained from the sun. In this area, solar radiation models are an extremely valuable piece of equipment to have.
2. The monthly average daily global solar radiation is estimated with the use of the modified Angstrom-type equation in most solar radiation models.
3. It is feasible to draw the conclusion that the models that have been supplied in this work may be utilized for the estimate of solar radiation at a specific site and perhaps in other areas with climatic parameters that are like those found in the original location.

4.4.6 Solar Radiation Data Processing (SRDP)

In the present work, better techniques for estimating solar radiations are proposed for use in circumstances when such radiations are not measured according to standards or for operational reasons. Following the presentation of the methods for processing the data of solar radiation datasets gathered from various sources in the sections that came before this one, it is now essential to present the methods for processing the data of solar radiation datasets collected from numerous sources. The reason for this is that the models that are developed are based on a variety of characteristics, each of which varies throughout a large range of topography and geography. This chapter provides a description of the database, the locations that are being investigated, as well as the tools that are being used to analyse the data.

Geographical Knowledge for the study region

The preferred locations are located across the district of Kalaburagi in the Indian state of Karnataka. In Fig. 4.10, all the locations are plotted on a map, and Table 4.2 lists the geographical and statistical information for each site. Because the Indian Meteorological Department did not have data for all the places, data were obtained from a variety of accessible sources (due to the nonexistence of observing stations in the experimental area). All data for the research region were provided daily, including solar radiation (MJ/m^2), temperature values (°C), precipitation (mm), wind (m/s), and relative humidity (fraction). The locations mentioned above include not just a broad spectrum of longitudes and latitudes in the northern hemisphere, but also a broad spectrum of temperature ranges and surface features. These locations include tropical, subtropical, temperate, and frigid climates.

Data Processing Techniques

The first thing that must be done with the data in question is processing it. It is essential, not only to assemble datasets in a manner that is user-friendly, but also to get rid of any numbers that are incorrect or missing. Data sets spanning a variety of values were collected from selected sites around the study area. They were screened and processed to produce data sets that were consistent and tidy. This task was completed to generate daily data for their evaluations. In this case, a quick evaluation of the daily data set assimilation might be beneficial. By adding up the daily radiation provided by all sites, monthly radiation results were computed.

Quality Assurance

When one takes a deeper look at the data sets that are supplied by meteorological stations, one will often find that there is a lack of data that is of good quality, sometimes over extended periods. This is because of inaccurate measurements, which are brought on by a confluence of variables such as the inaccuracy and unpredictability of the equipment, as well as problems and flaws in the operating processes. The database in issue is not an exception; like other databases, it must first go through the filtering process before it can be utilized for modelling. The method of quality control that was based on a series of physical tests and a semi-automated statistical test was used for each dataset to impact the likely removal of erroneous data that was brought on by measurement uncertainties. It was created primarily to undertake data quality checks and, if necessary, to identify and eradicate any false data values.

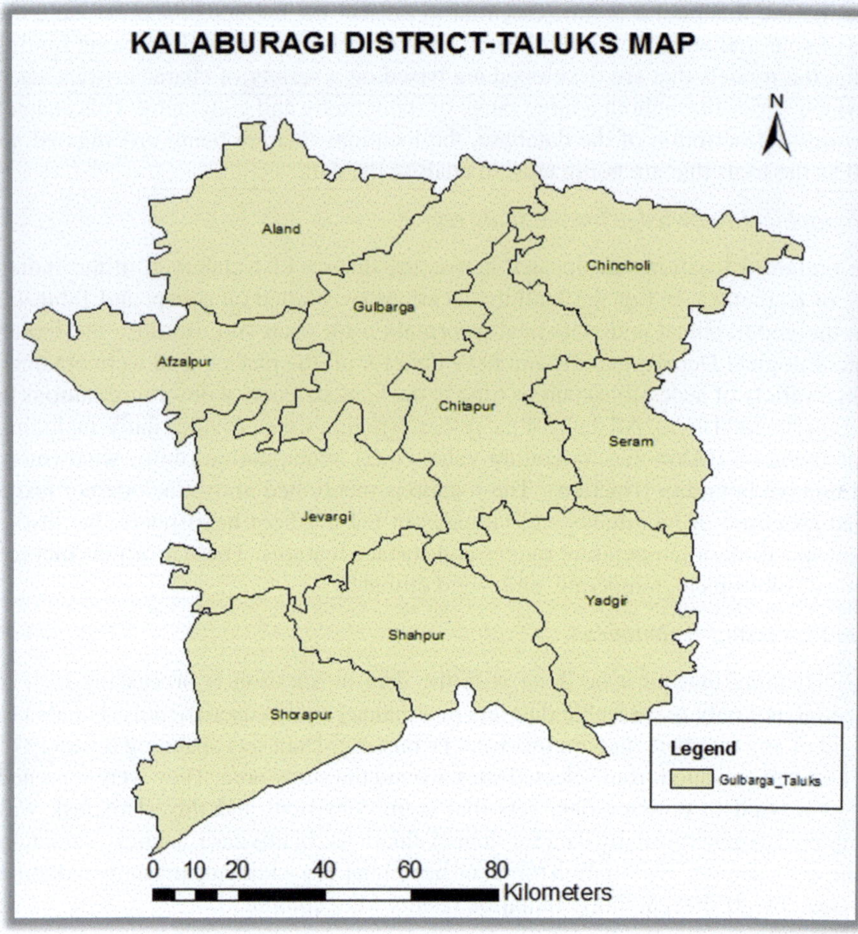

Fig. 4.10 Location of sites on the map

Assimilation

The compilation of daily data sets is the subject of this portion of the chapter. Original data provided by numerous stations are used for this purpose. The quality control technique was not applied to daily data sets. Radiation that is measured regularly is not more likely to provide erroneous results, and the daily data assist in adjusting for measurement errors. The data for all four Indian sites were sourced from open-source datasets such as NREL, TAMU, and RET screen. All the data for all the sites, including radiation values, temperature, precipitation, wind speed, and relative humidity values, were submitted daily.

The difficulty of the entire sun-to-electricity pipeline necessitates a multidisciplinary approach to properly address the entire process. However, due to the need

Table 4.2 Values of solar radiation at different locations

Sl. No	Geographic Location's				
	Place	Latitude(°N)	Longitude(°E)	Solar Insolation	Elevation (m)
1	Afzalpur	17° 14′ 42.988″ N	76° 24′ 8.161″ E	18.902 MJ/m^2	424 m
2	Aland	17° 32′ 8.658″ N	76° 35′ 40.831″ E	18.700 MJ/m^2	516 m
3	Chincholi	17° 27′ 14.048″ N	77° 20′ 30.861″ E	18.925 MJ/m^2	469 m
4	Chitapur	17° 8′ 57.874″ N	77° 3′ 2.712″ E	18.894 MJ/m^2	444 m
5	Gulbarga	17° 21′ 53.704″ N	76° 51′ 48.539″ E	18.882 MJ/m^2	468 m
6	Jevargi	16° 56′ 7.706″ N	76° 37′ 57.585" E	18.986 MJ/m^2	400 m
7	Seram	17° 4′ 32.733″ N	77° 18′ 57.506″ E	18.976 MJ/m^2	602 m
8	Shahpur	16° 39′ 25.366" N	76° 55′ 52.907″ E	16.702 MJ/m^2	407 m
9	Shorapur	16° 29′ 16.062″ N	76° 33′ 48.884″ E	19.117 MJ/m^2	474 m
10	Yadgir	16° 43′ 56.165″ N	77° 15′ 40.292″ E	19.296 MJ/m^2	412 m

for a significant quantity of data from several sources, including detailed geometric models of the buildings and the surrounding environment and meteorological data series, this procedure is challenging. To produce finer datasets by processing the data obtained from spatial radiation models and to create a valuable dataset, the datasets can be checked, sorted, summarized, aggregated, analysed, reported, and categorized. Missing data imputation methods are also included in the processing techniques, which range from manual to automatic to electronic data processing procedures. The whole sequence of actions necessary for the data processing technique is depicted in Fig. 4.11. The data on insolation, global solar radiation and other meteorological variables should be neatly arranged. The primary goal of data assimilation is to determine solar radiation periods and the spatial distribution of solar irradiance to perform spatial interpolation and visualization.

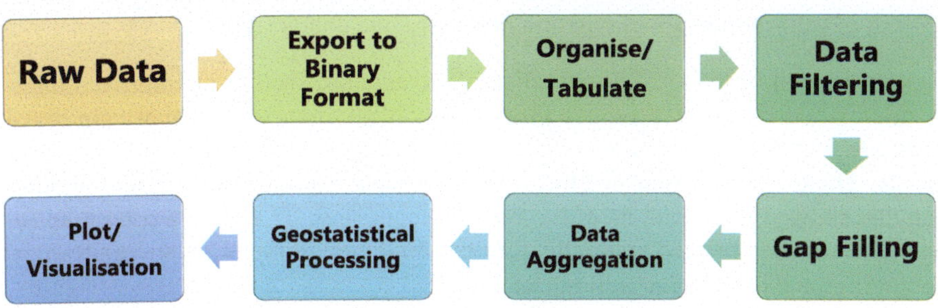

Fig. 4.11 Flowchart illustrating data processing techniques

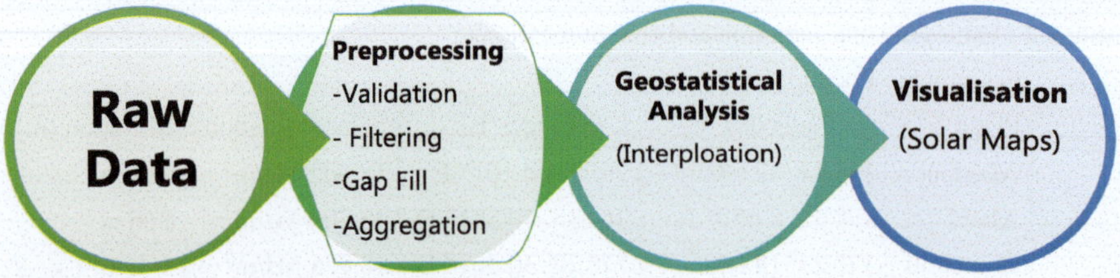

Fig. 4.12 Illustration for solar map generation techniques

4.4.7 Solar Map Generation

By using a suitable cartographic representation of solar insolation characteristics, it is possible to get a better knowledge of the geographical distribution of solar potential, PV output, air temperature, and any other spatial aspect. The information depicted on maps was incomparable to that which could be gleaned from incomprehensible tables of figures or complicated graphs consisting of lines, bars, and other elements.

The procedures necessary to build solar maps from raw data are depicted in Fig. 4.12. The solar map is a set of mapping tools that have never been seen before, and they provide significant benefits to those in the solar energy field:

1. A system for use by real estate developers, property scouts, and energy experts; its purpose is to locate the most suitable locations for solar energy projects.
2. Instructions for administrators, technical staff, and analysts: comprehensive quantitative analysis of the area's solar potential, presented in a spatially detailed format.
3. For managers and marketers: Marketing tools that are helpful and eye-catching maps that are interesting to look at are a great way to boost the exposure and presentation of your brand.
4. (Self) Educators: A map is the ideal instructional tool since it elucidates a great deal of information about climatic processes and characteristics.

Apart from it, it may result in a range of advantages, such as an improved interpretation of solar resource ranges in a geographical context, data that help make decisions, and business arguments presented via eye-catching maps with attractive visuals. There was one map for the whole year, one for each month, and another map that showed the locations of the measurement sites. The sample solar map (in Fig. 4.13) being generated using the elaborated method in Fig. 4.12 is visualized in the findings and discussion section (Sect 4.5).

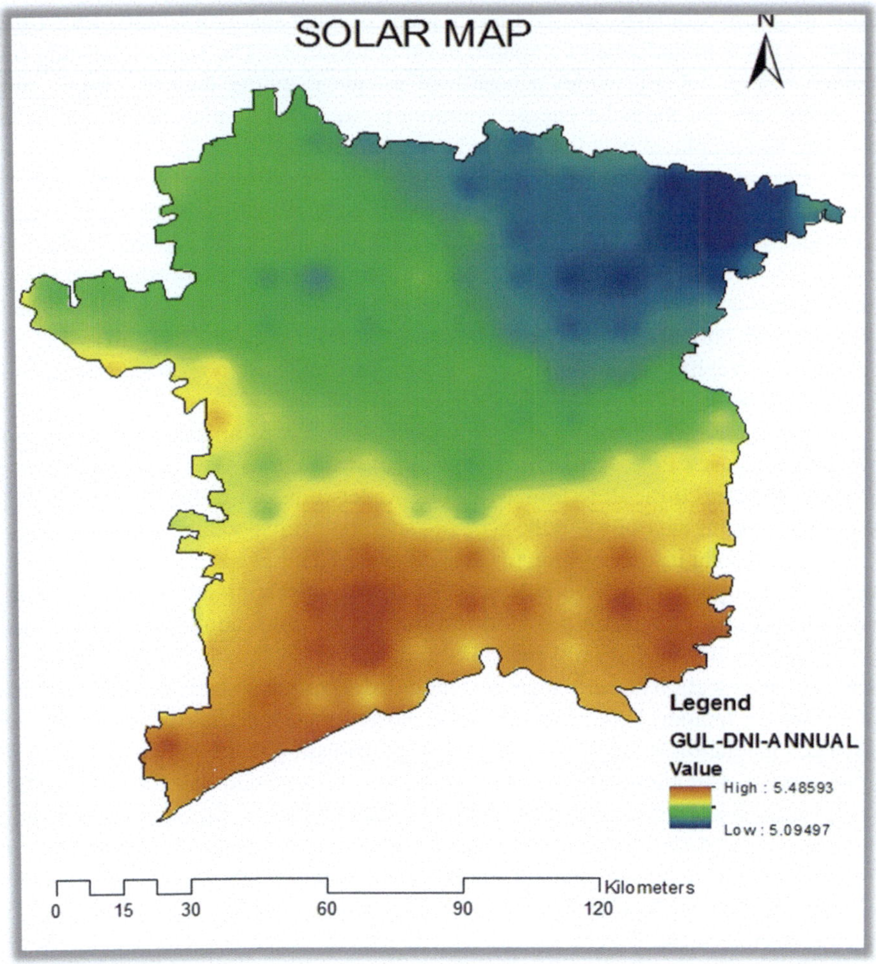

Fig. 4.13 Illustration for solar insolation map

4.5 Results and Discussions

The transmission of energy between the surface and the atmosphere results in incident solar radiation that reaches the planet's surface. Recent months have seen a significant increase in the amount of effort made towards the creation of exact, physical interpretations of solar radiation across topography and other surfaces (such as plant canopies).

Figure 4.13 makes it easier to locate the distribution and availability of solar resources in the Kalaburagi district. The southern part of the district has more solar potential, whereas the northern-eastern part has lower solar insolation levels, as seen in this diagram. Similarly, the average amount of solar insolation statistics in the central region is mixed when compared to other locations. The same solar map (as depicted in Fig. 4.13) provides a wealth of unique mapping data that might pay off handsomely for the solar energy community, as it can display accessible solar resources in a graphical way that anybody can understand.

The capacity to execute such models has been aided by the development of potent analytic tools as part of geospatial technology, which also offers an excellent modelling environment for the interface of inputs and outputs of solar radiation models. These models may be used to engineering, agriculture, forestry, conservation, and management, as well as hydrologic and ecological modelling. For a broad range of applications, the capacity to simulate the solar energy that is absorbed by complex topographic surfaces is necessary. The implementation is progressing well, and most of the theory is now in place.

The only goal should not be to get increasingly better estimations of actual sun radiation. Because of the demand for more input data, the models become increasingly complex and difficult to utilize. Future models must eliminate this stumbling block by providing flexibility in the type of radiation estimated and the data input.

The annual solar insolation map for the Kalaburagi district is depicted in Fig. 4.14. Even at the taluk level, the figure depicts annual sun insolation numbers. It demonstrates that annual global sun insolation values in Chincholi taluka were lower, but annual global solar insolation values in Shorapur and Shahpur were greater. Similarly, the centre section of the map, which included Afzalpur and the majority regions of Gulbarga, Chitapur, Seram, and Jevargi, showed a normal distribution of yearly global sun insolation values. These annual solar insolation maps were able to reveal the resource distribution pattern within the region. These could be extremely useful for all forms of solar resource planning and implementation. Similarly, depending on the project's needs, maps might be created on a monthly or weekly basis. Which will allow us to visually evaluate information with care for a variety of applications. We may also prepare downscaled maps ranging from the global to the local level to meet our various needs.

4.6 Summary

Solar energy resources have become one of the primary movers in the energy market, as we all know. To start any solar energy project anywhere on the planet, we should be aware of the resources available at that location. We normally use field-based or in-situ measurement and satellite-based measurement of solar resources using tools like radiometers to estimate the quantity of solar insolation or radiation available at any site. However, because of technological breakthroughs, well-established atmospheric models can now be used to determine the precise quantity of solar insolation at any location on the planet. Also, because radiation measurements are affluent, there is a clear need not only to develop a precise solar radiation model for use in various time stamps that can account for solar resource measurement using various

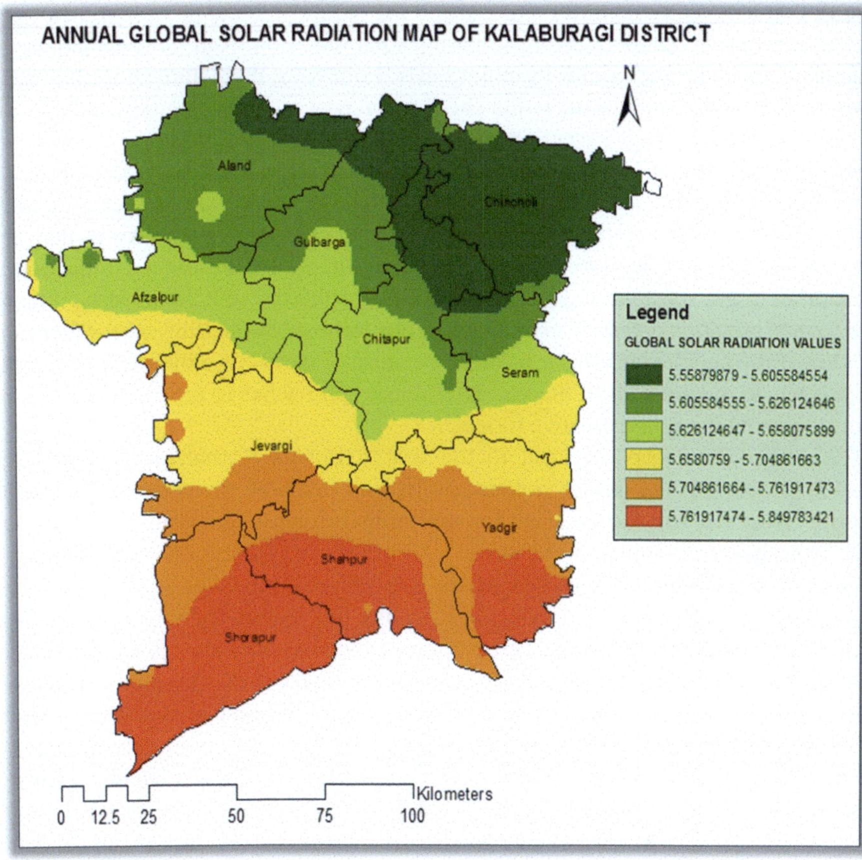

Fig. 4.14 Annual solar insolation map of Kalaburagi district

parameters such as Day of Year (DOY), latitude, topographic parameters, and radiation parameters, but there is also a clear need to develop a precise solar radiation model for use in various time stamps that can account for solar resource measurement using various parameters such as Day of Year (DOY), latitude, topographic parameters, and radiation parameters. In a similar vein, the current research aimed to create a model for the Indian environment that could predict the quantity of global solar irradiation on an area and location basis. The same models, on the other hand, can be used to estimate other aspects of solar radiation, such as diffuse and direct radiation.

The current part of the study tried to create model-to-model solar energy over the experimental study areas. This involved taking into consideration several factors, including (i) the estimation of existing or prospective solar radiation, (ii) the existing technology for generating energy from incoming solar radiation, and (iii) the portion of earth that is most suitable for exposure to the sun. The focus is on the development of non-traditional models for estimating solar radiation from satellite data both on an area scale and on a point-by-point basis. Following that, statistical measurements such as the mean bias deviation (MBD) and root mean square deviation were used to assess the models' overall performance (RMSD). In the end, a brand-new comparison measure that was given the name "accuracy score" was developed to estimate the effectiveness of the model. The precision score was merged with the "sum deviation," which is the product of the MBD and RMSD measures of dispersion. Several different satellite-derived solar radiation estimations were compared to one another to get substantial divergence results. Furthermore, it demonstrated how open to interpretation a variety of statistical approaches to modelling solar radiation are. One of the conclusions that was made because of that is an exhaustive research analysis has to integrate all of the key assessment approaches to produce a criterion for constructing the most effective model.

The designed concept was put to the test in India's Karnataka state, in the Kalaburagi district. The model yielded results that followed the same well-known patterns. We must use the datasets for multidimensional characteristics once the modelling phase is completed. As a result, to make use of these datasets, we must first process them using various procedures to preserve all the data's attributes. Not only is it essential to combine datasets in a manner that is user-friendly, but it is also essential to eliminate any missing, erroneous, or inconsistent values throughout the data processing of data obtained utilizing modelling approaches. This is an extremely important step. Detailed data processing strategies are introduced and summarized in this section of the study.

Moreover, a proper cartographic depiction of processed solar insolation measurements was necessary to gain a better knowledge of solar potential spatial distribution patterns and other geographical characteristics. In this type of setting, an appropriate schema was developed to plot the solar maps to expose knowledge that can be interpreted. This type of information cannot be gleaned from unreadable tables that are typically filled with numbers or from standard graphs that are cluttered with lines, bars, and other features. Solar maps provided a collection of one-of-a-kind mapping technologies to give a more effective solution to these problems. As a result, the solar energy community may anticipate receiving a sizeable financial reward for their efforts. Aside from that, it paves the way for a deeper knowledge of the diversity of solar resources in a geographical context, as well as data that may support judgements and be used for commercial purposes. One annual map and twelve-monthly maps were created to test the solar map production methodologies.

References

Abujayyab SKM, Almotairi KH, Alswaitti M, Amr SSA, Alkarkhi AFM, Taşoğlu E, Hussein AM (2021) Effects of meteorological parameters on surface water loss in burdur lake, turkey over 34 years landsat google earth engine time-series. Land 10(12). https://doi.org/10.3390/land10121301

Agugiaro G, Remondino F, Stevanato G, Filippi RD, Furlanello C, Kessler FB, Modelling U, Unit E (2005) Estimation of solar radiation on building roofs in mountainous areas. Int Arch Photogrammetry Remote Sens Spat Inf Sci 38:155–160

Ahmad T, Thaheem MJ (2018) Economic sustainability assessment of residential buildings: a dedicated assessment framework and implications for BIM. Sustain Cities Soc 38(January):476–491. https://doi.org/10.1016/j.scs.2018.01.035

Ajayi O, Ohijeagbon O, Ajanaku K, Aasa S, Omotosho O (2016) Techno-economic assessment of renewable hybrid systems for rural electrification and distributed generation in selected sites across Nigeria. J Afr Res Bus Technol 2015:1–31. https://doi.org/10.5171/2015.124767

Alrwashdeh SS (2018) Assessment of photovoltaic energy production at different locations in Jordan. Int J Renew Energy Res 8(2):797–804

Anderson MC, Norman JM, Kustas WP, Houborg R, Starks PJ, Agam N (2008) A thermal-based remote sensing technique for routine mapping of land-surface carbon, water and energy fluxes from field to regional scales. Remote Sens Environ 112(12):4227–4241

Arefin R (2020) Groundwater potential zone identification using an analytic hierarchy process in Dhaka City, Bangladesh. Environ Earth Sci 79(11). https://doi.org/10.1007/s12665-020-09024-0

Barron M G, Vivian D, Yee SH, Diamond SA (2008) Temporal and spatial variation in solar radiation and photoenhanced toxicity risks of spilled oil in Prince William Sound, Alaska, USA. Environ Toxicol Chem/SETAC 27(3):727–736. https://doi.org/10.1897/07-317

Bechtel B, Alexander PJ, Böhner J, Ching J, Conrad O, Feddema J, Mills G, See L, Stewart I (2015) Mapping local climate zones for a worldwide database of the form and function of cities 4(1):199–219. https://doi.org/10.3390/ijgi4010199

Borfecchia F, Caiaffa E, Pollino M, Cecco LD (2014) Remote sensing and GIS in planning photovoltaic potential of urban areas. Eur J Remote Sens 47(195):195–216. https://doi.org/10.5721/EuJRS20144713

Bornatico R, Hüssy J, Witzig A, Guzzella L (2013) Surrogate modeling for the fast optimization of energy systems. Energy 57:653–662. https://doi.org/10.1016/j.energy.2013.05.044

Chrysoulakis N (2002) Energy in the urban environment: Use of terra/aster imagery as a tool in urban planning. J Indian Soc Remote Sens 30(4):245–254. https://doi.org/10.1007/BF03000367

Chu T, Guo X (2013) Remote sensing techniques in monitoring post-fire effects and patterns of forest recovery in boreal forest regions: a review. Remote Sens. https://doi.org/10.3390/rs6010470

Chwieduk DA (2009) Recommendation on modelling of solar energy incident on a building envelope. Renew Energy 34(3):736–741. https://doi.org/10.1016/j.renene.2008.04.005

Dawson DA, Vercruysse K, Wright N (2020) A spatial framework to explore needs and opportunities for interoperable urban flood management. Philosophical transactions of the royal society a: mathematical, physical and engineering sciences, vol 378, no 2168. https://doi.org/10.1098/rsta.2019.0205

Delkash M, Al-Faraj FAM, Scholz M (2018) Impacts of Anthropogenic land use changes on nutrient concentrations in surface waterbodies: a review. Clean Soil Air Water 46(5). https://doi.org/10.1002/clen.201800051

Dubayah R, Rich PM (1996) GIS-based solar radiation modeling. GIS Environ Modeling 129–134

Fadare DA (2009) Modelling of solar energy potential in Nigeria using an artificial neural network model. Appl Energy 86(9):1410–1422. https://doi.org/10.1016/j.apenergy.2008.12.005

Gambhir A, Dixit S, Toro V, Singh V, Ramachandra T, Jain V, Krishnadas R, Nayak G, Prajapati JK, Noel Ja, Cressie, Christopher KW, Lean JL, Sayigh AAM, Pareek A, Studios B, Commission CE, National T, Plan A, Change C, Nehru J, Observatory N (2010) A guide to PV system design and installation. Solar Energy, vol 44, no 5. Academic Press. https://doi.org/10.1002/wcc.018

Gómez-Chova L, Amorós-López J, Mateo-García G, Muñoz-Marí J, Camps-Valls G (2017) Cloud masking and removal in remote sensing image time series. J Appl Remote Sens 11(1). https://doi.org/10.1117/1.JRS.11.015005

Goodchild MF (2009) Geographic information systems and science: today and tomorrow. Proc Earth Planetary Sci 1(1):1037–1043. https://doi.org/10.1016/j.proeps.2009.09.160

Hassan QK, Rahman KM, Haque AS, Ali A (2011) Solar energy modelling over a residential community in the City of Calgary, Alberta, Canada. Int J Photoenergy. https://doi.org/10.1155/2011/216519

Hwang JT, Liu HC (2008) A study on cadastral maps data changes processing and display. In: 29th Asian conference on remote sensing 2008, ACRS 2008, vol 3, pp 1658–1663. https://www.scopus.com/inward/record.uri?eid=2-s2.0-84865637957&partnerID=40&md5=286d099ecf3cf2552ea839fdfc098b36

Khandelwal S, Goyal R, Kaul N, Mathew A (2016) Assessment of land surface temperature variation due to change in elevation of area surrounding Jaipur, India. Egypt J Remote Sens Space Sci 1–8. https://doi.org/10.1016/j.ejrs.2017.01.005

Khatib T, Mohamed A, Sopian K (2012) A review of solar energy modeling techniques. Renew Sustain Energy Rev 16(5):2864–2869. https://doi.org/10.1016/j.rser.2012.01.064

Khorrami B, Heidarlou HB, Feizizadeh B (2021) Evaluation of the environmental impacts of urbanization from the viewpoint of increased skin temperatures: a case study from Istanbul. Turkey Appl Geomat 13(3):311–324. https://doi.org/10.1007/s12518-020-00350-3

Kumar A, Gomathinayagam S, Giridhar G, Mitra I, Vashistha R, Meyer R, Schwandt M, Chhatbar K, Delhi N, Gmbh S (2008) Solar resource assessment and mapping of India. SolarPaces 2012, Imd

Kumar D (2019) Hyper-temporal variability analysis of solar insolation with respect to local seasons. Remote Sens Appl Society Environ 15:100241. https://doi.org/10.1016/j.rsase.2019.100241

Kumar D (2020) Satellite-based solar energy potential analysis for southern states of India. Energy Rep 6:1487–1500. https://doi.org/10.1016/j.egyr.2020.05.028

Kumar R, Bhattacharya B, Kumar SVVA, Chakraborty A, Bhowmick SA, Sharma R, Sharma S (2017) Harnessing non-conventional energy potential using space borne sensors. In: 38th Asian conference on remote sensing—space applications: touching human lives, ACRS 2017, 2017-Oct

Li D, Wu B, Chen B, Qin C, Wang Y, Zhang Y, Xue Y (2020) Open-surface river extraction based on sentinel-2 MSI imagery and DEM Data: case study of the upper yellow river. Remote Sens 12(17). https://doi.org/10.3390/RS12172737

Liu Y (2009) Modelling Urban development with geographical information systems and cellular automata. In: Transition, p 188. http://www.crcpress.com

Loukanov A, El Allaoui N, Omor A, Elmadani FZ, Bouayad K, Seiichiro N, Lachowycz K, Jones AP, Houlden V, Porto de Albuquerque J, Weich S, Jarvis S, Davies C, Lafortezza R, Boulton C, Dedekorkut-Howes A, Holden M, Byrne JA, Chan IYS, He HS (2020) Effects of neighborhood building density, height, greenspace, and cleanliness on indoor environment and health of building occupants. Environ Res 106(February):213–222. https://doi.org/10.1016/j.buildenv.2018.06.028

Luo H, Paal SG (2021) Metaheuristic least squares support vector machine-based lateral strength modelling of reinforced concrete columns subjected to earthquake loads. Structures 33:748–758. https://doi.org/10.1016/j.istruc.2021.04.048

Meisen P, Quéneudec E, Yuan M, Nara A, Bothwell J, Ramirez L, Zink R, Dorner W, Stoeglehner G, Sharma P, Harinarayana T, Journal E, Studies E, Vol M, Division ES, Jacoby HD, Supervisor T, Marks DH, Systems E, Ramkrishna B (2006) Overview of renewable energy potential of India. Modeling Earth Syst Environ 2(October):1–20. https://doi.org/10.1016/j.compenvurbsys.2015.03.002

Mhawej M, Nasrallah A, Abunnasr Y, Fadel A, Faour G (2021) Better irrigation management using the satellite-based adjusted single crop coefficient (aKc) for over sixty crop types in California, USA. Agric Water Manage 256. https://doi.org/10.1016/j.agwat.2021.107059

Myers DR (2005) Solar radiation modeling and measurements for renewable energy applications: data and model quality. Energy 30(9 SPEC. ISS.):1517–1531. https://doi.org/10.1016/j.energy.2004.04.034

Narimani MR, Molzahn DK, Crow ML (2021) Tightening QC relaxations of AC optimal power flow problems via complex per unit normalization. IEEE Trans Power Syst 36(1):281–291. https://doi.org/10.1109/TPWRS.2020.3004289

Ostapowicz K (2009) Solar Energy Modelling with ArcGIS

Padmanaban R, Bhowmik AK, Cabral P (2019) Satellite image fusion to detect changing surface permeability and emerging urban heat islands in a fast-growing city 14(1). https://doi.org/10.1371/journal.pone.0208949

Peng J, Hu Y, Liu Y, Ma J, Zhao S (2018) A new approach for urban-rural fringe identification: Integrating impervious surface area and spatial continuous wavelet transform. Landscape Urban Plann 175(May 2017):72–79. https://doi.org/10.1016/j.landurbplan.2018.03.008

Rao M, Jaiswal RK, Singh P, Raghawaswamy V, Pathan SK, Gurumukhi KT (2005) A national urban information system - EO and GIS concept. In: International astronautical federation—56th international astronautical congress 2005, vol 2, pp 1345–1365. https://www.scopus.com/inward/record.uri?eid=2-s2.0-34250833227&partnerID=40&md5=02c7011acde3bebcfb475273339d864b

Rao P, Gupta K, Roy A, Balan R (2021) Spatio-temporal analysis of land surface temperature for identification of heat wave risk and vulnerability hotspots in Indo-Gangetic Plains of India. Theoret Appl Climatol 146(1–2):567–582. https://doi.org/10.1007/s00704-021-03756-0

Rn HK, Shafiullah GM, Maung A, Oo T, Stojcevski A (2014) Feasibility study of standalone hybrid power system modeled with Photovoltaic modules and Ethanol generator for Victoria. Int J Eng Sci Invention 3(8):54–69

Santos T, Gomes N, Freire S, Brito MC, Santos L, Tenedório JA (2014) Applications of solar mapping in the urban environment. Appl Geogr 51:48–57. https://doi.org/10.1016/j.apgeog.2014.03.008

Sawle Y, Gupta SC, Bohre AK (2016) PV-wind hybrid system: a review with case study. Cogent Eng 3(1):1–31. https://doi.org/10.1080/23311916.2016.1189305

Sumari NS, Cobbinah PB, Ujoh F, Xu G (2020) On the absurdity of rapid urbanization: spatio-temporal analysis of land-use changes in Morogoro, Tanzania. Cities 107 https://doi.org/10.1016/j.cities.2020.102876

Tajbakhsh A, Shamsi A (2019) Sustainability performance of countries matters: a non-parametric index. J Clean Prod 224:506–522. https://doi.org/10.1016/j.jclepro.2019.03.189

Tanskanen A, Jantunen A, Saksa JM, Partanen J, Bergman J, Bertsch V, Treitz M, Geldermann J, Rentz O, Di Silvio B, Cesarotti V, Introna V, Babatunde OM, Babatunde DE, Denwigwe IH, Adedoja TB, Adedoja OS, Okharedia TE, Kwakwa PA, Authors F (2021) New directions in UK energy policy. Int J Energy Sect Manage 15(2):157–172. https://doi.org/10.1108/IJESM-12-2019-0001

Velotto D, Bentes C, Tings B, Lehner S (2016) First comparison of Sentinel-1 and TerraSAR-X data in the framework of maritime targets detection: South Italy case. IEEE J Oceanic Eng 41(4):993–1006. https://doi.org/10.1109/JOE.2016.2520216

Vinod Kumar R, Singh SK (2018) Solar photovoltaic modeling and simulation: as a renewable energy solution. Energy Rep 4:701–712. https://doi.org/10.1016/j.egyr.2018.09.008

Wahlström Å, Joosen S, Geurts F, Kleefkens O, Zavrl MS, Hansen K, Norvaisa E, Makowska N, Despretz H, Ruud S (2008) Handbook for performing feasibility studies of alternative energy systems. November, pp 1–93. https://ec.europa.eu/energy/intelligent/projects/sites/iee-projects/files/projects/documents/sentro_handbook_en.pdf

Wang J, Han P, Li G (2021) The extraction and analysis of deformation features of Hualien earthquake based on sentinel-1A/B SAR data. J Indian Soc Remote Sens 49(9):2069–2077. https://doi.org/10.1007/s12524-021-01377-8

Weng Q, Hu X (2008) Medium spatial resolution satellite imagery for estimating and mapping urban impervious surfaces using LSMA and ANN. IEEE Trans Geosci Remote Sens 46(8):2397–2406. https://doi.org/10.1109/TGRS.2008.917601

Yang F, Chen W, Zhang J, Sun X (2021) Follow the perturbed approximate leader for solving semi-bandit combinatorial optimization. Frontiers Comput Sci 15(5). https://doi.org/10.1007/s11704-020-9519-9

Yang HC, Chang CM, Urban PL (2019) Automation of fizzy extraction enabled by inexpensive open-source modules. Heliyon 5(5):e01639. https://doi.org/10.1016/j.heliyon.2019.e01639

Yang J, Huang X, Tang Q (2020). Satellite-derived river width and its spatiotemporal patterns in China during 1990–2015. Remote Sens Environ 247. https://doi.org/10.1016/j.rse.2020.111918

Zhao J, Liu H, Dong R (2008) Sustainable urban development: Policy framework for sustainable consumption and production. Int J Sustain Dev World 15(4):318–325. https://doi.org/10.3843/SusDev.15.4:5a

Chapter 5
Quantifying the Amount of Energy Harvest

Abstract Investors, consumers, and even suppliers are worried about PV technology. Few studies have estimated solar insolation annually. When building a PV system, it's important to determine the quantity of accessible sunlight. Sun radiation is measured by solar radiance and insolation. Solar radiance is measured in kW/m^2. Solar radiance ranges from 0 to 1 kW/m^2 throughout the day. Location and weather affect solar brightness. Daily global and/or direct radiation measurements determine sun radiance, pyranometers, and/or pyrheliometers monitor global radiation (measuring direct radiation). This information comes from well-established sources accumulated over two decades. Solar insolation is more typically employed in system design than solar irradiance. Sun insolation is the total solar energy received at a particular place during a certain period (kW/m^2 /day). Solar insolation and solar irradiance both measure power density (for solar insolation, "hours" in the numerator and "day" in the denominator are time measurements), but solar insolation is instantaneous solar irradiance averaged over a particular period. Solar insolation data are utilized for basic PV system design, whereas solar radiance data are used for sophisticated PV system performance calculations. Solar insolation may be represented in several units than MJ/m^2/year.

Keywords *Solar insolation · Usable area · Photovoltaic · Pointwise and area-wise solar insolation*

5.1 Overview

In countries that have made legal promises that are both explicit and reliable, the use of solar power produced by photovoltaics (PV) is starting to become more widespread in the energy market. Solar photovoltaic (PV) technology is quickly becoming one of the most promising technologies for the generation of renewable energy on-site (Pramanick et al. 2021; Rey-Mahía et al. 2022). The adoption of solar technology is rising at a fast rate, and suitable surfaces in urban and rural regions are playing a crucial role for both private and public investors, as well as the inhabitants of such places. In addition to topography, the conditions of the atmosphere (such as air turbidity, cloudiness, aerosol, and water vapour, for example) have a role in

D. Kumar, *Renewable Energy Scenarios in Future Indian Smart Cities*, Advances in Geographical and Environmental Sciences, https://doi.org/10.1007/978-981-19-8456-3_5

the accuracy of the computation of the amount of incoming solar radiation (e.g., shadowing by hills, mountains, nearby buildings, vegetation, etc.). In situations with no clouds in the sky, an exact solar insolation calculation may be made using a digital elevation model. When there is fog in the sky, the dependability of the system decreases because there are more unknowns in the parameterization of the processes and situations that occur in the atmosphere. As a result, after a certain point of the estimation process, models must employ experimental data to adjust estimates to real values (Buyantuyev and Wu 2010; Mbow et al. 2014; Wang et al. 2021; Yu et al. 2014). World atmospheric data centres routinely calculate and deliver global atmospheric data reflecting these constituents with spatial resolutions ranging from around 1 to 500 km. The accuracy of the clear-sky irradiance computation is mainly dependent on knowledge concerning aerosols. In a nutshell, this chapter will begin by outlining the criteria and process for selecting the best software tools for performing solar analysis (Tarigan et al. 2014). After that, there will be an overview of the general workflow as well as full descriptions of each phase and theory used to create this methodology.

5.2 Sub-objectives

(1) Quantifying the accessible amount of solar insolation for energy harvest.
(2) Potential photovoltaic (PV) yield estimations.

5.3 Workflow Overview

Figure 5.1 depicts the comprehensive workflow for the methodology used in this chapter. The project has two parts: the first will assess the quantity of solar insolation that may be harvested for energy, and the second will estimate the possible photovoltaic yields. The study's main goal is to calculate the amount of solar insolation. To meet the needs of various users, the amount of solar insolation has been calculated in both an area and a pointwise manner.

These main steps summarise the tasks involved in (1) obtaining and manipulating the required dataset, (2) determining the area-wise incident global solar radiation, (3) determining the pointwise global incident solar radiation, and (4) determining the appropriate surface area for PV deployment, as well as (5) determining the total solar PV electricity estimates.

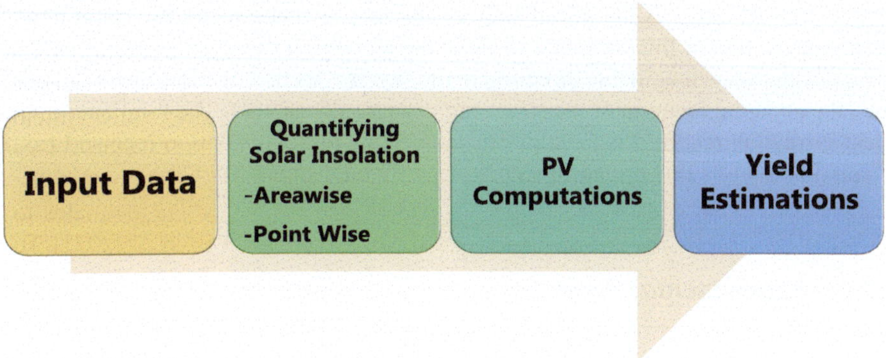

Fig. 5.1 Schematic for workflow diagram

5.4 Methods

This section covers the approach, data sets, and software/tools utilized to accomplish the goal. The current section of the project employs both quantitative and analytical research methodologies. Methodologies for analytical and quantitative research often involve aspects such as the estimate of values and the selection of processes to utilize in data analysis. The analytical design is an essential component of the study design since it enables us to pick the method by which the data will be assessed, in addition to providing information on the outcomes of the data/observation and the product. Quantitative research techniques make use of a wide array of mathematical models and statistical algorithms to arrive at the results that are needed. Analysing the situation and gaining an understanding of the connection between information that has been visually interpreted or digitally processed and the real state of affairs or phenomena that is taking place on the ground is also an analytical method. To quantify the values of a certain parameter, quantitative procedures are also necessary.

Numerical models were used to calculate solar insolation, which was parameterized by a set of inputs describing the locations, state of the atmosphere, and topographical characteristics. There are a lot of different variables and processes that impact the quantity of solar insolation that is received at a certain location on the surface of the globe. The rapid computation of clear-sky irradiance from a set of input parameters is made possible by this technique, which has been the subject of investigation in several different research. One of the predictable parameters that may be represented properly by numerical models is the location of the sun. At the local scale, the amount of incoming solar radiation is affected by variables such as the geography of the surrounding area, albedo, and the forest canopy. Strong local insolation gradients are formed because of variations in elevation, surface orientation (slope and aspect), and shadows produced by topographic features. For each place on the DEM, the solar analyst builds an upward-looking hemispheric viewshed (virtual fisheye photography) as the initial step. A sun map is made using the same hemispheric projection as the viewshed to show the amount of direct solar radiation that arrives from every imaginable direction in the sky. It does this by plotting, over some time, the apparent location of the sun using discrete sky sectors. The total amount of direct insolation received is equal to the sum of the direct insolation received from all sun map sectors that are unimpeded (or partly unobstructed) by the viewshed.

5.4.1 Experimental Area

Because it exhibits characteristics of a typical urban setting, the Central University of Karnataka (CUK) campus on Aland Road, Kadaganchi, Karnataka State, India, was chosen as the case study area. It is in the heart of the district, at a position with a low population density, but it has large electrical needs and buildings with a wide range of morphologies and characteristics, including age, height, and function. This area has a wide range of qualities that can be examined for solar energy potential. The university planning office provided the data for the master plan of the campus. The University administrative buildings were asked to provide a whole square block of data. Structures that were not included in the study were eliminated unless they had a major shadowing effect or were near any of the buildings. Figure 5.2 is an illustration of the geographical setting of the study region Figs. (5.3, 5.4 and 5.4)

5.4.2 Datasets Used

The GCPs were acquired using a handheld GPS, a 1:50 K scale Survey of India topographic sheet, SRTM DEM/BHUVAN DEM, and very high-resolution (VHR) satellite data from Google Earth (site validations). Other open archive data sets for execution and validation from BHUVAN NOEDA (NRSC), Landsat Archive & Earth Explorer (USGS). The data were acquired from a variety of open-source or educational websites, as well as on-the-ground data. Additional information was gathered from books, journals, reports, and websites. The following are some of the many information sources mentioned:

(1) Online gateways and databases *(ResearchGate)*.
(2) Ministry of New and Renewable Energy *(Website, Reports, Manuals)*.
(3) Online Journals *(e.g., Solar Energy, Energy, Journal of Solar Energy Engineering)*.
(4) Books in the context of infrastructure management, trends in energy sciences.
(5) Reference texts from ESRI, ERDAS Inc. publications, and manuals.

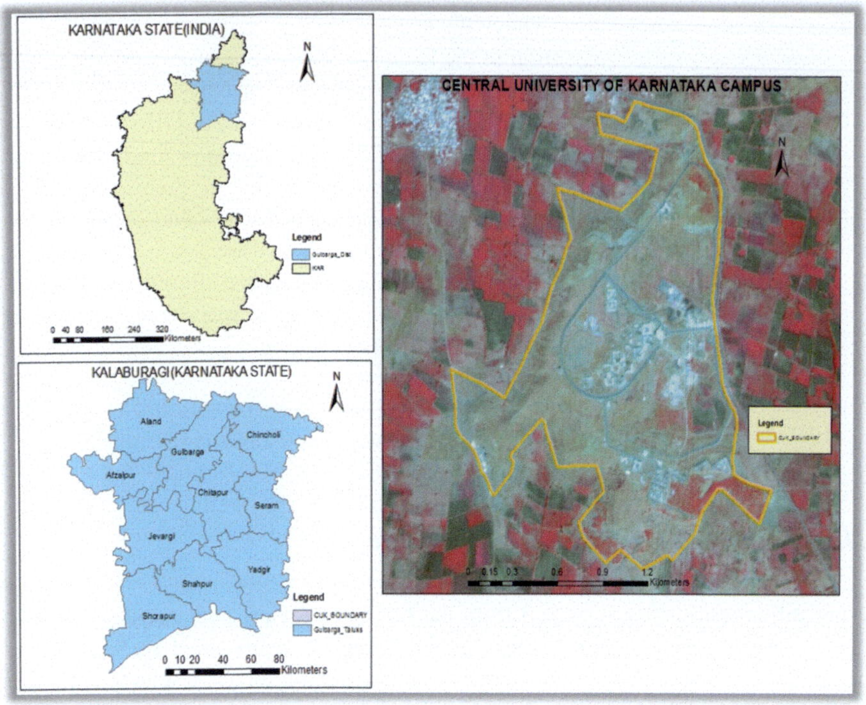

Fig. 5.2 Experimental area location

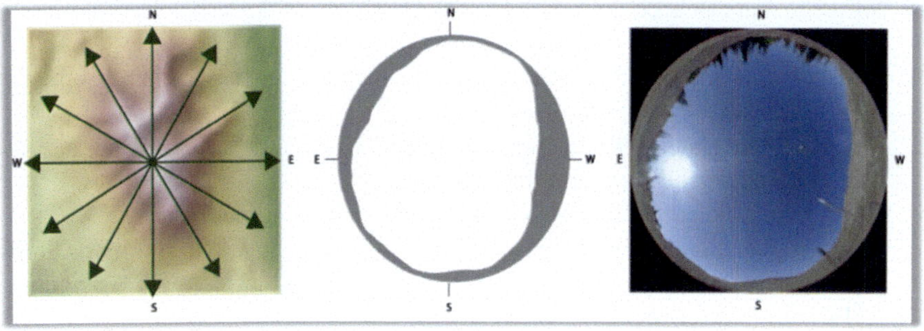

Fig. 5.3 Diagram depicting the horizontal angles, the resulting viewshed, and the viewshed superimposed over the sky view

5.4.3 *Software and Tools Used*

(1) Energy Plus 8.0
(2) Geospatial Toolkit *(NREL)*
(3) System Advisor Model *(SAM)*

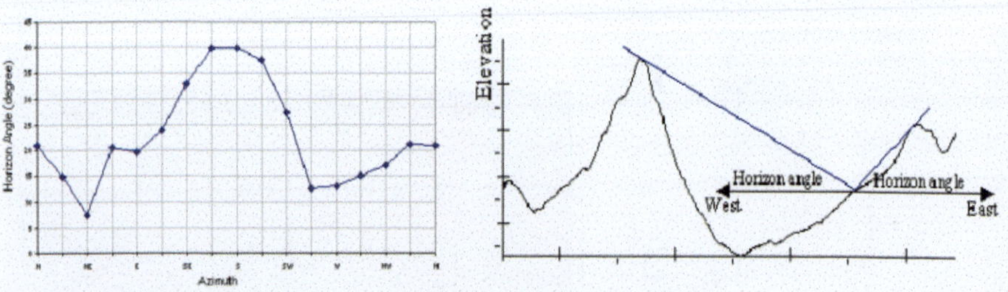

Fig. 5.4 Illustration for the horizontal angles (degree) versus Azimuth

(4) RET Screen Plus Software
(5) ESRI ArcGIS 10. x (*for vector data analysis*)
(6) ERDAS 201x (*for image processing*).

5.4.4 Software Selection Criteria Process

A few software tools were picked from the literature review and assessed in depth after much deliberation and analysis of the criteria (as discussed in Sect. 2.9 of chapter 2). The following criteria were explored to aid the software selection process in determining viable solar software solutions for the current study work:

(1) A programme that could analyse complex data containing various parameters and zones.
(2) Software that can account for solar radiation losses owing to barriers that create shading.
(3) A software product with a built-in weather database or the capacity to import weather data from a variety of sources.
(4) A software package with a user-friendly interface, attractive visuals, and no need for specialized programming abilities.
(5) A software application that is extremely time and location sensitive.
(6) Software that can precisely measure the gross surface area of building rooftops and occupied areas by undesired items (such as HVACs, green roofs, pipelines, and so on) to discover acceptable PV application sites.
(7) Software that does not require a lot of processing power.

5.4.5 Area-Wise Solar Insolation

The ArcMap extension tool version 10x. solar analyst performs an analysis of the amount of available sunlight by taking into consideration the site's latitude, the topography, the amount of shadow cast, and the air's attenuation. It gives information on direct, diffuse, and global irradiation, as well as direct irradiation time, submaps, sky maps, and viewsheds. In particular, the solar analyst model calls for direct atmospheric transmissivity along the route with the least amount of atmospheric resistance. Radiation may be broken down into two categories: direct and diffuse. Direct radiation accounts for the bulk of the total radiation, while diffuse radiation comes in second. Except in areas surrounded by highly reflective surfaces, such as snow cover, the fraction of total radiation that is accounted for by reflected radiation is rather small. Spatial analyst's solar radiation tools do not include reflected radiation in their total radiation calculations. Because of this, the amount of total radiation may be found by combining the amounts of direct and diffuse radiation. With the help of solar radiation tools, one may make calculations for specific points or whole geographical regions. This process includes the following four steps:

(1) The estimation of an upward-looking hemispheric viewshed involves the use of topography.
(2) Place the viewshed on top of a map of direct sun to get an estimate of direct radiation.
(3) Estimate the amount of diffuse radiation by superimposing the viewshed on a sky map that depicts diffuse conditions.
(4) To generate an insolation map, you must first repeat step 4 for each location of interest in the map.

The creation of an upward-looking hemispheric viewshed for each point in the digital elevation model is an essential part of the process. This is necessary because topography and surface characteristics may have a significant impact on the radiation that is emitted (DEM). Except in areas surrounded by highly reflective surfaces, such as snow cover, the fraction of total radiation that is accounted for by reflected radiation is rather small. Spatial analyst's solar radiation tools do not include reflected radiation in their total radiation calculations. There is a correlation between the quantity of visible sky and the amount of insolation that a location receives. A sensor that is placed in an open field, for instance, will get a greater amount of insolation than one that is placed in a steep ravine. The picture that may be seen below is an upward-looking hemispheric shot that depicts the visible sky as well as the sky directions that are veiled by the neighbouring terrain and surface features. This is the perspective that one would have if one were standing on the ground and looking up in every direction.

Calculation of the viewshed

From a certain perspective point, the viewshed is a raster representation of the whole sky and how much of it is visible or obscured. To calculate a viewshed, one must first determine the greatest angle of sky blockage, also known as the horizon angle, in a certain number of directions all around a place of interest. Interpolation is used to determine the horizontal angles for all other directions that have not been searched. Following this step, the horizon angles are converted to a hemispherical coordinate system. This produces a two-dimensional raster picture that depicts a three-dimensional globe of different orientations. Each raster pixel in the viewshed is given a value that matches whether the sky direction can be seen from that pixel or if it is obscured by anything else. On the globe of directions, the positions of the output cells (row and column) link to the zenith angle, which is the angle related to the direction of straight above, and the azimuth angle, which is the angle relative to the direction of north.

The following figure illustrates how the computation of a viewshed for one cell of a DEM is done. A hemispheric depiction of the sky is generated by first computing the angles of the horizon and then selecting a predetermined number of directions to employ in the process. The viewshed that is created because of this process decides whether or not sky directions may be seen (shown in white) or if they are hidden (shown in black) (shown in grey). A hemispheric picture has the viewshed overlaid on it so that the idea may be shown.

In conjunction with information regarding the position of the sun and the direction in which the sky is facing, viewsheds are utilized as a tool for the calculation of direct, diffuse, and total (direct + diffuse) radiation for all locations, as well as for the production of an accurate insolation map (represented by a sun map and sky map, respectively)

Sun map calculation

When calculating the quantity of direct solar radiation that comes from each sky direction, a sun map that has the same hemispheric projection as the viewshed is used. A raster depiction of the path or apparent location of the sun as it moves across the sky during the day and year is what's known as a sun map. This is analogous to gazing up at the sky and seeing how the position of the sun changes over time. The location of the sun at various periods throughout the day (in terms of both hours and days) and throughout the year gives rise to the formation of distinct sectors on the sun map (days or months).

To compute sun tracking, you will need to know the latitude of the area under study as well as the time configuration that will define the sun map sectors. Each submap sector is provided with its one-of-a-kind identification value, in addition to the centroid's zenith angle and azimuth angle. It is determined on an individual basis how much solar radiation each sector receives, and then the viewshed is placed over the sun map to determine how much direct radiation is received.

The sun's path through the sky as seen from 45° north latitude from the winter solstice (December 21) to the summer solstice (June 21) is shown in Fig. 5.5. (June 21). (June 21). The position of the sun is shown throughout the day in intervals of half an hour and monthly intervals during the year in each solar sector, which is represented by a coloured box. It is important to point out that the picture is projected in the same hemispheric plane as viewsheds that are directed upward. At various periods throughout the day and the year, the path that the sun takes across the sky is illustrated here (Figs. 5.6 and 5.7).

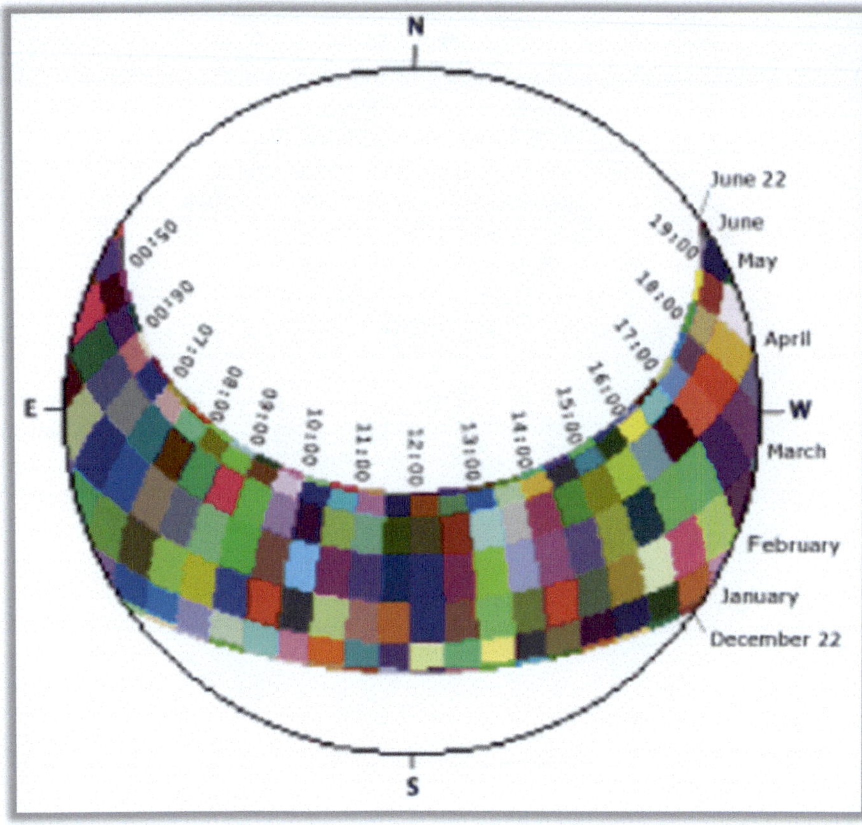

Fig. 5.5 Illustration for sun map example

Sky map computation

Diffuse radiation is electromagnetic energy that is emitted from all directions in the sky. This kind of radiation is produced when particles in the air scatter the radiation (clouds, particles, and so forth). To calculate diffuse radiation for a particular location, a sky map is constructed to depict a hemispheric view of the whole sky separated into a series of sky sectors defined by zenith and azimuth angles. Each sky sector represents a different portion of the sky. Each sector is given its distinctive distinguishing value, in addition to the centroid's zenith and azimuth angles. It is possible to calculate the diffuse radiation emitted by each sky sector depending on the direction it faces (zenith and azimuth). In the following sky map, eight zenith divisions and sixteen azimuth divisions have been used to describe the different sky sectors. Each colour represents a separate sky sector, often known as a region of the sky, from which diffuse radiation is emitted.

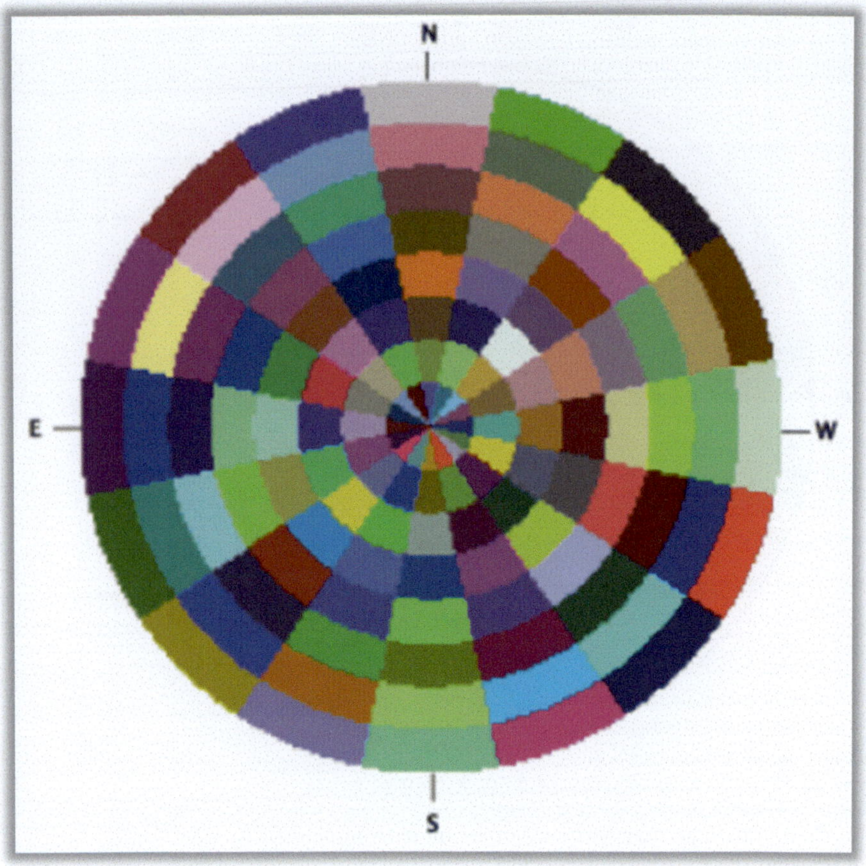

Fig. 5.6 Illustration for sky map example

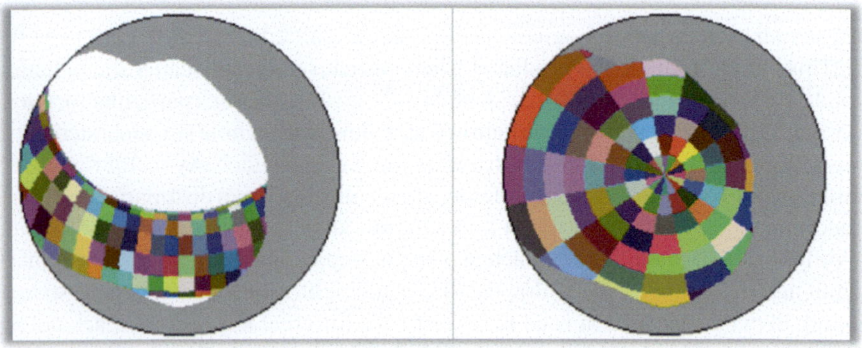

Fig. 5.7 Illustration for the overlaid viewshed with the sun map example (on the left) and the overlaid viewshed with the sky map example (on the right)

Viewshed Overlay with sun map and sky map

A raster viewshed is used in conjunction with a raster sun map and a raster sky map to calculate insolation. Thus, the amount of diffuse and direct radiation from all directions may be calculated. The percentage of area that can be seen above ground in each sector may be calculated by taking the total number of cells in that sector and dividing it by the number of unobstructed cells. Consideration is given to areas of the sky that are just partially obscured. The following graphic illustrates how a viewshed may be overlaid on top of a sun map as well as a sky map. When referring to the sky, the colour grey indicates that some directions are obscured. When calculating solar radiation, the total amount of direct and diffuse insolation that comes from directions with an unobstructed sky is added together.

Usage of area-wise solar radiation computations

(1) Because calculating insolation might take a significant amount of time, it is essential to verify and recheck that all the values are accurate. Digital elevation models (DEMs) may be calculated in a matter of hours for smaller areas but might take days for larger ones.
(2) The output radiation raster's will always be of the floating-point type, and the units they will use are watt-hours per square metre (Wh/m^2). In the case of the direct duration raster, the output is an integer that may be interpreted as the number of hours.
(3) To figure out things like the sun's declination and where exactly it is in the sky, we need to know the site's latitude. Latitude is measured in decimal degrees, with positive numbers indicating locations in the northern hemisphere and negative numbers indicating locations in the southern hemisphere.
(4) Since the study is often restricted to regionally relevant landscape sizes, it is generally suitable to use a single latitude value for the whole DEM. Larger datasets, such as those for states, countries, or continents, might provide results that differ greatly from those obtained by measuring insolation at a single latitude (greater than 1 degree). To examine broader geographical areas, it is important to split the study area into zones at varying latitudes.

5.4.6 Pointwise Solar Insolation

Incoming solar radiation is the primary factor behind all of earth's physical and biological processes (insolation). The ability to calculate the quantity of insolation at a given place is useful in a variety of industries, including agriculture, resource management, meteorology, civil engineering, and ecological study. The following processes are included in calculations for point locations or the full geography (shown in Fig. 5.8).

For example, knowing how much sun an area gets over time could be beneficial in deciding where to build a new ski resort or where to cultivate specialty crops that require specific microclimate conditions for maximum growth. Maps showing

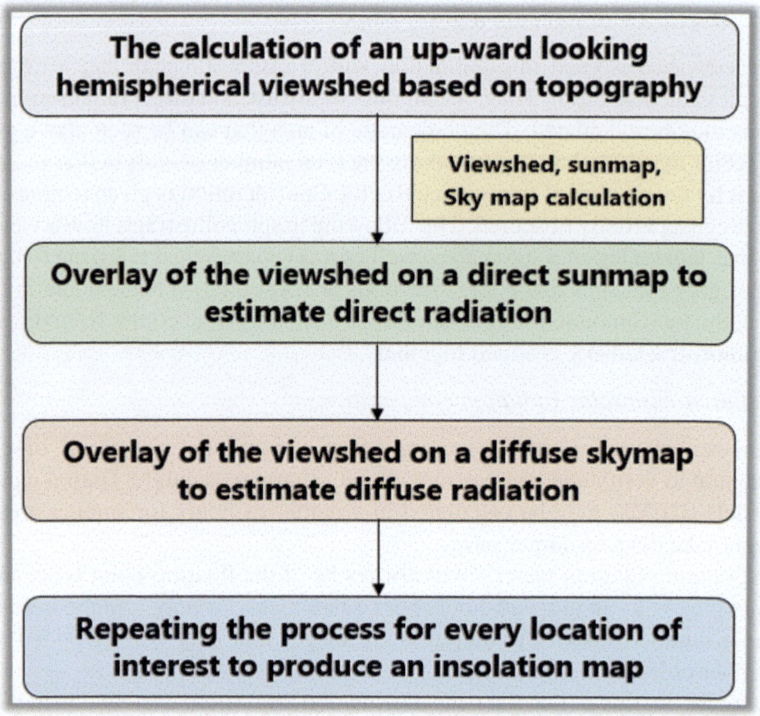

Fig. 5.8 Schematic showing pointwise solar insolation computations

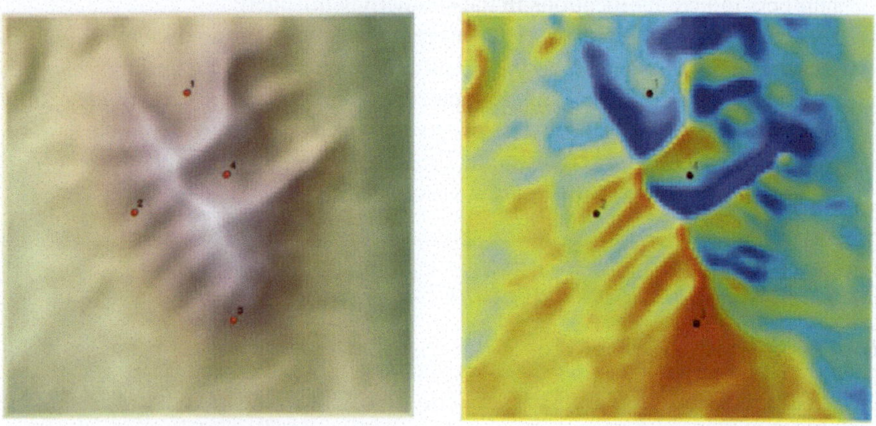

(a) Sample locations on a DEM to analyze solar radiation

(b) Sample radiation values for the entire area

Fig. 5.9 Illustrations for pointwise solar insolation computations

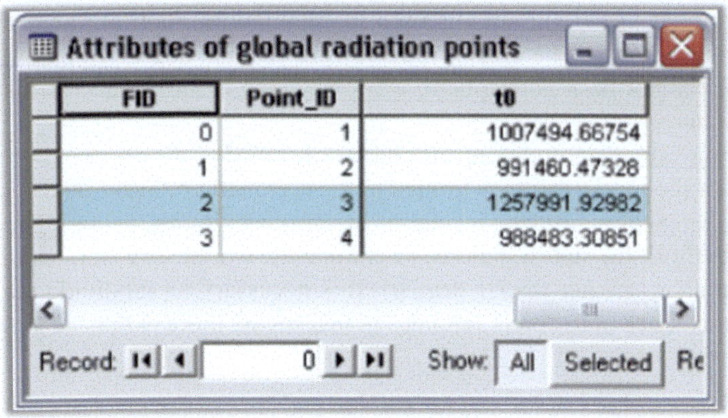

Attributes of global radiation points

FID	Point_ID	t0
0	1	1007494.66754
1	2	991460.47328
2	3	1257991.92982
3	4	988483.30851

Fig. 5.10 Illustration showing attribute generated after pointwise solar insolation computation

insolation, for instance, have been shown to help predict the behaviour of forest fires and come to conclusions about the most effective methods of firefighting. In the fields of civil engineering and urban planning, insolation may be an important factor in suitability models, which help determine the most advantageous places to build. Four spots (numbered red points) along a hillside have been selected as suitable vineyard plots in the following graphic. It must be decided which site will receive the most solar exposure during the growing season (April–October) to enhance crop growth (Figs. 5.9 and 5.10).

The worldwide insolation (direct diffuse, Wh/m^2) for the entire research region has been estimated using the point solar radiation analysis, indicating where the biggest levels of radiation are throughout the summer months (red = high insolation; blue = low insolation). The attribute table shows the calculated global insolation for each site location. Based on this criterion, Location 3 (highlighted in blue) has the most insolation and may be the best location to produce grapes.

More in-depth analyses may be able to zero in on an ideal range of insolation, also known as a regime, about the time of day during which insolation is received and the proportion of direct to diffuse radiation. These results might readily be incorporated into more complicated algorithms to make predictions about the areas of a region that are most suited for grape production.

Usage of pointwise solar radiation computations

(1) Because calculating insolation might take a significant amount of time, it is essential to verify and recheck that all of the values are accurate. The calculation of a large digital elevation model (DEM) may take several hours, whereas a very large DEM can take many days.
(2) The output radiation raster's will always be of the floating-point type, and the units they will use are watt-hours per square metre (Wh/m^2). The direct duration raster will produce an integer that represents the number of hours as its result.

(3) The latitude of the site area is used in calculations such as solar declination and solar position. The units for latitude are decimal degrees; positive latitudes are for the northern hemisphere, while negative latitudes are for the southern hemisphere.
(4) Utilizing a single latitude value for the whole DEM is often appropriate since the scope of the investigation is limited to the regionally relevant landscape scales. The findings of measuring insolation at various latitudes, using bigger datasets such as those for states, nations, or continents, would vary tremendously from one another (greater than 1 degree). It is necessary to subdivide the research area into zones of different latitudes to investigate more extensive geographical regions.

5.4.7 Solar Energy Estimations

Regardless of season, temperature, geography, or topography, the sun provides 1000 W/m^2 of solar insolation throughout the year. We may use the following mathematical calculation to calculate the area required to install a solar panel that can create 1 kWh of electricity (which is the minimum requirement for any normal household energy usage) and consider 24 h of sunshine.

$$\frac{1\ \text{kWh}}{(1000\ \text{W/m}^2) * 24\ \text{h}} = 0.04\ \text{m}^2 \tag{5.1}$$

However, we can view the solar energy setup components such as the battery, inverter/converter, and wiring strings, as well as the generation configuration of a terrain, sunshine, and seasonal temperatures. We will conclude that obtaining steady or uniform solar insolation throughout the day and seasons has numerous limitations. As a result, the sun will not shine for 24 h (at most, it will shine for 8–9 h), so factor in 8 h. There will also be some losses during the energy storage in the battery, lowering the battery efficiency to roughly 75%. (Which will affect the photovoltaic energy availability due reduction in efficiency concerning time). Similarly, the DC to AC conversion process causes losses in the system, resulting in a 75% fall in efficiency (in the optimal case). In general, any solar system's maximum efficiency remains around 30%, but actual efficiency ranges from 8 to 18%, therefore 8% may be regarded as an excellent scenario. After considering all of the system's limitations, Eq. 5.1 will be rewritten as

$$\frac{1\ \text{kWh}}{(1000\ \text{W/m}^2) * (8\ \text{h}) * (0.75) * (0.75) * (0.08)} \sim 3\ \text{m}^2 \tag{5.2}$$

The above Eq. 5.2 states and provides a thumb rule for photovoltaic systems that the least area required to generate 1kwh of solar energy is 3 m^2.

5.4.8 *Photovoltaic Yield Calculations*

The availability of DEM via remotely sensed data and the development of GIS have ushered in a revolution in the development of distributed models to explain meteorological and geographical spatial phenomena, particularly in areas where there are few or no conventional meteorological observatories, such as in rugged and remote areas.

Calculating the Effective Roof Area

The remote sensing image is classified into two classes using ISO-data clustering techniques: I roof area; and (ii) others. The roof surfaces facing south and southwest are identified because they were able to gather the most quantity of incident solar radiation in the northern hemisphere. To figure out which rooftops were suitable for the installation of photovoltaic panels, it was necessary to have a mechanism that could calculate the total surface area of each rooftop. The gross area was determined by digitizing the campus of the research area using ESRI ArcGIS 10 and making use of the data that are publicly accessible on Google Earth. The geographic information system (GIS) was chosen because it is software that enables powerful changes to be achieved on geographic data and offers a practical technique for the investigation and image creation of geographic data. It also contains a rich collection of analysis tools. The 5 m resolution orthophotos from 2013 were utilized in this thesis because they were the highest-resolution pictures available on Google Earth at the time. To build a new layer with the extracted rooftops, a Google Earth image can be digitized using GIS editing tools. The compute geometry tool within the attribute tables can then be used to determine the digitized rooftop surface areas. The gross surface area estimates were more accurate as a result of this technique. After determining the entire gross surface area, the following step was to calculate the total area that may be used for PV installation. The following are two approaches for evaluating usable rooftop areas. Below, we'll go further into the two primary methodologies behind these correction factors: (a) best-published practices for assessing useable rooftop areas employing PV corrective factors, and (b) digitally digitalized obstacles utilizing GIS on roofs decreased by a PV module coverage factor.

Usable Area—*PV Corrective Factors*

A growing body of research literature has emerged in recent years to aid in the evaluation of rooftop areas suitable for PV installations. This collection of articles explores methods for accurately calculating the amount of available land appropriate for the installation of PV systems by using correction variables known as PV access factors. The percentage of rooftop PV that is typically accessible varies across the various categories of common building types. Population intensity and build-up density, together with building types like residential and industrial, are two major groups identified in the literature. This typical building type was determined by calculating the density of the region of interest relative to the total number of people or structures. It is commonly known that as urban populations swell, the amount of roof space available per person must inevitably decline. Therefore, in this thesis, factors such as construction density or population density were ignored. The reason for this is that it is difficult to determine the quantity of available roof space for PV deployment since the experimental project is located mostly in an administrative building of the Central University of Karnataka. Also, this method shines in examinations of whole cities or regions as opposed to single structures. Therefore, we only considered PV access factors that had been calculated for a specific kind of building. In the research, several kinds of buildings were used to categorize PV access variables. Commercial and residential structures were identified as the two most common types of buildings. Because of their work, we now know PV access parameters for both cold and warm areas. Several significant barriers to accessing solar energy have been uncovered.

(1) **Roof Type**—Flat roofs and pitched roofs are the two most common varieties. The angle at which the PV systems may be mounted depends on this, thus it is crucial information to have.
(2) **Structural Adequacy**—Since PV systems increase the weight of rooftops, this factor is taken into consideration. For this reason, it is important to examine the roofs' structural soundness considering the building codes' criteria (wind loading, snow loading). It should be stressed that this is typically not a problem.
(3) **Material Compatibility**—When we talk about how nice something looks and how well it works for a PV system, we're talking about this. Unfortunately, this problem is hardly thought about.
(4) **Shading**—Reduced solar radiation on the roofs is taken into consideration here, since it may be produced by a variety of things, including trees, HVACs, other equipment, vents, chimneys, and other roof structures.
(5) **Orientation**—Which way the roof's surface is facing? If you have a flat roof, you won't have to worry about it, but it may have a major effect on a pitched roof.
(6) **Module Coverage**—This variable provides for the distance that must be left between photovoltaic (PV) modules, inverters, wiring, approach to modules, and other components for routine maintenance. This number was used as a "packing factor" to adjust the PV system's configuration and reduce its power density. To determine the amount of flat roof space accessible in commercial and industrial buildings in temperate regions, the following equation was discovered by (Chaudhari et al. 2018; Gao and Chaudhari, 2021).

$$\begin{aligned}\text{Usable Area} &= \text{Gross Area } \times \text{Material Compatibility}\\ &\times \text{Structural Adequacy } \times \text{Orientation}\\ &= \text{Gross Area} \times 100\% \times 100\% \times 65\% \times 100\% = 65\%\end{aligned} \tag{5.3}$$

This suggests that sixty-five per cent of the rooftop space has been determined to be acceptable for the installation of PV panels. After performing the computation, a packaging factor was used to limit the power density output of the PV array size, which allowed for the module coverage factor to be taken into consideration in their technique. In place of a percentage number, this technique relied on a ratio value to figure out how much electricity to draw from the PV system. Nevertheless, the percentages associated with this component were found in a variety of distinct sources. Two publications stood out because they each contributed a percentage value to the module coverage factor. This made both works extremely remarkable. Research on the photovoltaic (PV) access criterion for rooftop space was carried out in Austin, Texas, by (Sicard et al. 2016) based on the results. It was found that the module coverage factor is fifty per cent in a different study (Bergamasco and Asinari 2011), researchers looked at pitched residential and industrial roofs in Italy to determine the module coverage factor. They found that the factor was 45%. This indicates that 45 per cent of the total area may be used effectively. It was determined that both criteria were valid because neither value is affected by differences in climate or geographic location. A module exposure factor of 47.5% was decided by doing the necessary calculations to determine the mean of the two figures. After accounting for this component, the following calculation should be used to determine the proportion of the rooftop that can accommodate a PV installation:

$$\begin{aligned}\text{Usable Area} &= \text{Gross Area} \times \text{Material Compatibility} \\ &\quad \times \text{ Structural Adequacy} \times \text{Shading} \times \text{Orientation } \times \text{Module Coverage} \\ &= \text{Gross Area} \times 100\% \times 100\% \times 65\% \times 100\% \times 47.5\% \\ &= \text{Gross Area} \times 30.875\% \end{aligned} \tag{5.4}$$

This equation will be used for the results in Scenario 1.

Usable Area—*GIS Extracted Area and Module Coverage*

The second approach employed to assess useable rooftop space relied on the digitization of obstacles using GIS, the applicable PV correction factors, and the module coverage factor of 0.475. This method was used to determine whether the rooftop could be used for solar panels. The manual scanning and extraction of any vacant rooftop area that was filled by buildings such as HVACs, green roofs, chimneys, and other structures were completed to create a new layer. After that, we created a new layer in the GIS database called the "GIS extracted area" by combining the "gross area" of the rooftop with the "unusable zone" using the "erase" function in GIS. This GIS-retrieved area represents the portions of the rooftop that were vacant due to the absence of any buildings or other structures. After that, the recovered region from the GIS was applied to the coverage factor of 0.475 for the module. This would clear up sufficient outer space for the photovoltaic (PV) systems to be built and maintained accurately. The following equation was used during the investigation into Scenario 2.

$$\begin{aligned}\text{Usable Area} &= \text{GISExtracted Area } \times \text{Material Compatibility} \\ &\quad \times \text{ Structural Adequacy } \times \text{Orientation} \times \text{Module Coverage} \\ &= \text{GISExtracted Area} \times 100\% \times 100\% \times 100\% \times 47.5\% \\ &= \text{GISExtracted Area } \times 47.5\% \end{aligned} \tag{5.5}$$

The size of the PV system that should be installed on each building was determined by computing it based on the predicted useable area measurements for both cases.

In Figs. 5.11 a, b and c, we can see the different types of rooftops that are available for the study. These figures also show the methods for computing the rooftop area and finally locating the footprint of the building over the digital elevation model surface to visualize the mechanism of solar resource availability more efficiently and effectively. Figure 5.11b illustrates a straightforward rooftop configuration; but, in reality, the rooftop will be somewhat more complex, with a variety of elaborate mounting mechanisms. It may have a mast or mount structures in the centre or on the side, and because these structures will throw shadows on the incoming solar photovoltaic installation, it is recommended that they be removed or that the weight of this region is decreased. For accurate area estimates, we will need to use a variety of different formulas, which will vary according to the kind of rooftop and the accessibility of the structures located on the roofs.

5.5 Results and Discussions

The quantity of solar radiation that is intercepted at a particular location on the earth's surface remains affected not only by the surface's direction but also by the surrounding topography's ability sto obscure the sky. The number of sky barriers that may be seen from any particular place on or above complicated topographic terrain can be determined by utilizing viewshed models. An efficient method for rapidly determining the amount of solar radiation that has been intercepted has been developed by combining viewshed analysis with a lookup database that contains the irradiance from each direction of the sky. First, given the region of interest, the angular distribution of sky obstruction was determined and recorded in a hemispheric coordinate system. The total irradiance is then calculated by adding the values from the lookup tables for all clear-sky directions. A cosine correction was used to account for the angle of incidence. To estimate the geographical distribution of collected solar radiation, the operation was repeated throughout a topographic surface. Solar isolations or sky view factors were used to express the findings. The same viewshed approach can also be used to improve topographic surface reflected radiation calculations.

a

b

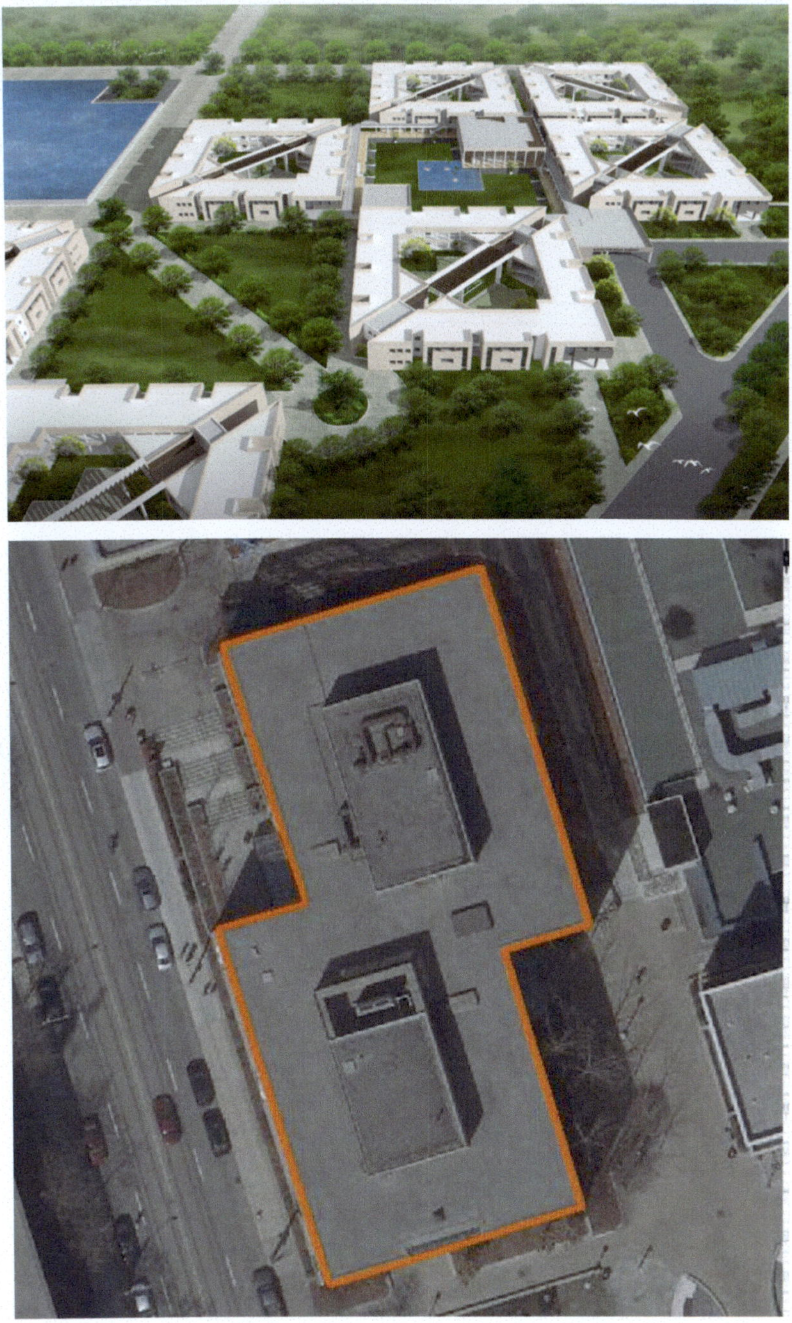

Fig. 5.11 **a** Sketch of available rooftop area over school buildings, **b** sketch for rooftop area estimation over a building, **c** Illustration showing building footprint over DEM surface

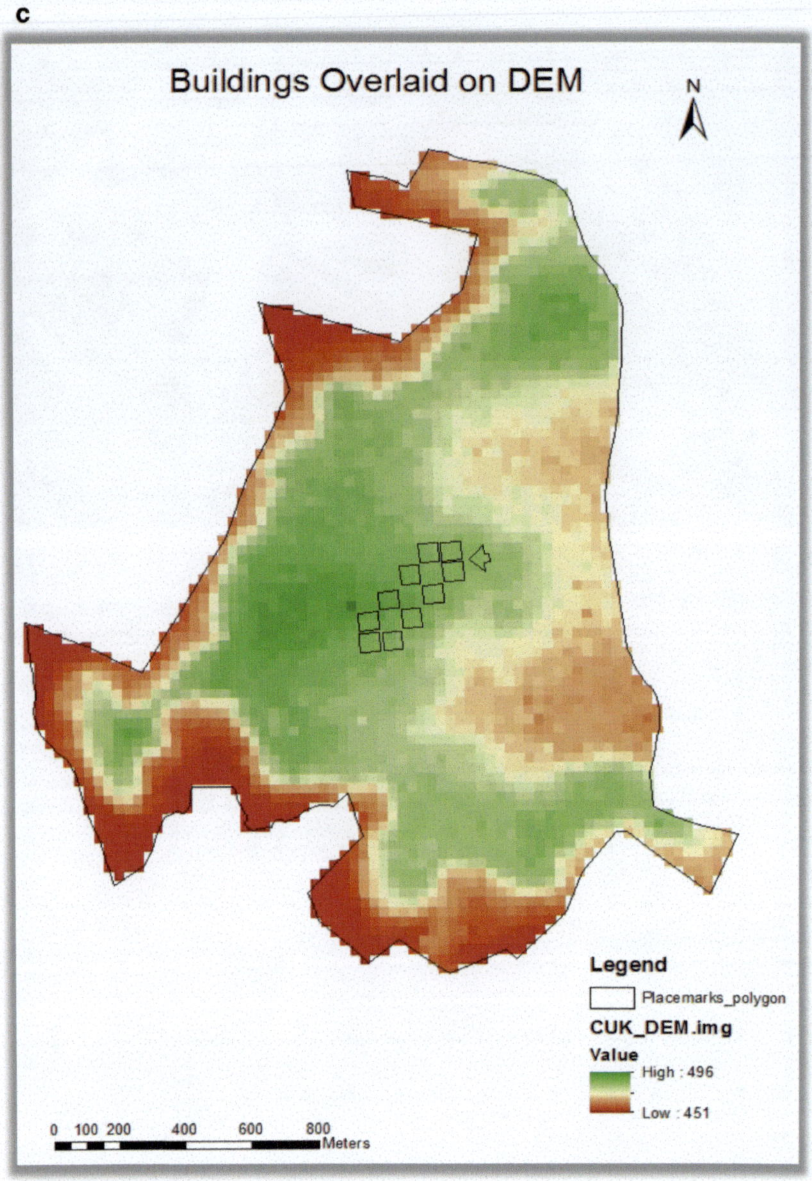

Fig. 5.11 (continued)

5.5.1 *Area-Wise Solar Insolation*

The quantity of solar insolation available at the Central University of Karnataka (CUK) Campus, which was used as a test region for pilot research, is depicted in Fig. 5.12. The graph in the illustration depicts the spatial distribution of available solar insolation (Wh/m^2/day) in the area depicted. It was calculated using BHUVAN (ISRO archive data) DEM version v3 with 30 m resolutions, sky size/resolution of 200, and time configuration of a whole year with a monthly interval of 1(one) hour for the year 2013, Z factor of 1, calculations directions of 32.

5.5.2 *Pointwise Solar Insolation*

The quantity of sun insolation accessible at several random sites throughout the Central University of Karnataka (CUK) Campus is depicted in Fig. 5.13. The figure depicts the spatial distribution of solar insolation (kWh/m^2/day) at various point locations within the exhibited area.

It was calculated using BHUVAN (ISRO archive data) DEM version v3 with 30 m resolution, sky size/resolution of 200, and time configuration of a whole year with a monthly interval of 1(one) hour for the year 2013, Z factor of 1, calculations directions of 32. It was given arbitrary sample marked places on campus as input points or tables. In Table 5.1, the computed pointwise values have been tabulated.

5.5.3 *Power Generation Calculations*

Estimating PV array size

Before beginning an accurate assessment of the PV array capacity (in kW) and surface area (in m^2) on the rooftop of each building, it is necessary to first choose the PV technology that will be used. Because there are so many different solar panel technologies available on the market today, the panel module that was deemed to be the most effective for the study area was selected based on its ability to meet the requirements for widespread implementation. Mono NeoN LG280N1C 280 W, SW 280 Silver Mono Plus, SL280CE-36P) and MULTICRSTALLINE PM 280 are the key panels now topping the market, according to markets and research. These panels were chosen based on their Ontario distributors' recommendations for cost, performance, and availability. The available surface area for panel installation has to be evaluated to establish how many panels could fit into the desired rooftop of a structure.

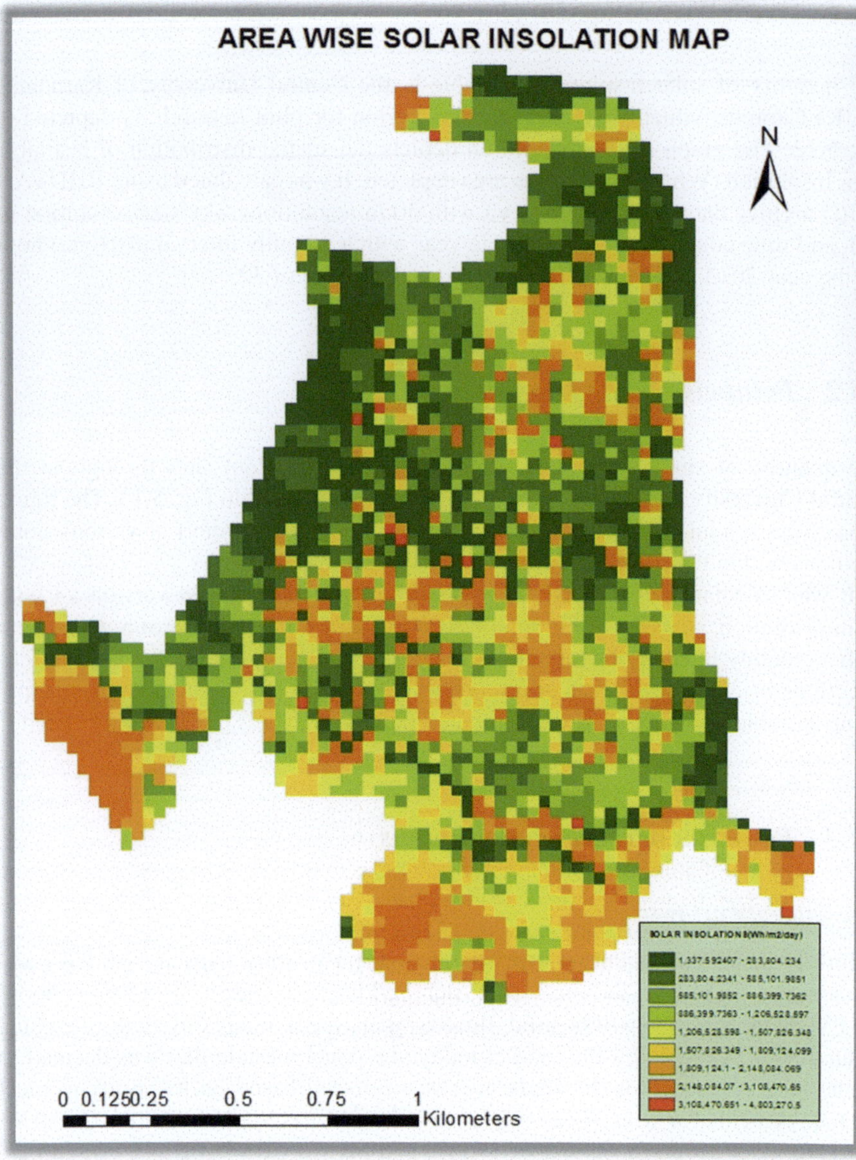

Fig. 5.12 Illustrations for area-wise annual insolation

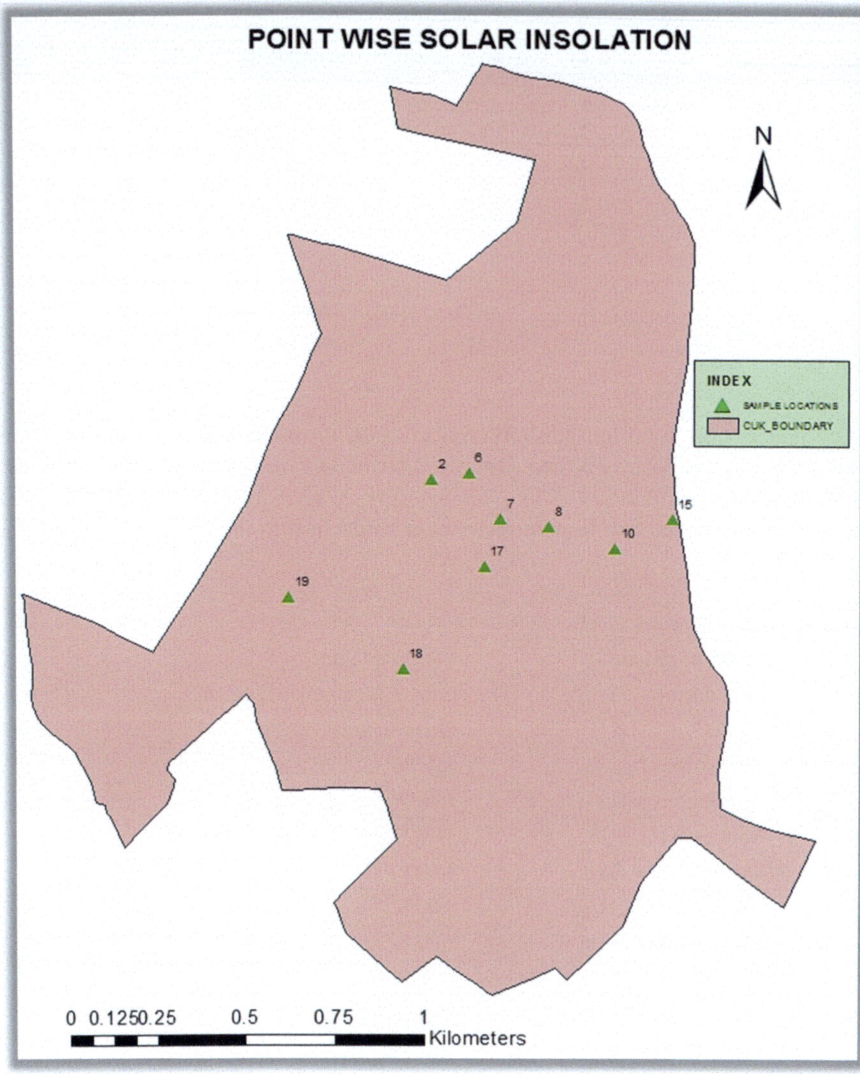

Fig. 5.13 Illustration for pointwise annual insolation

Before making an exact determination of the PV array capacity (in kW) and surface area (in m^2) on the rooftop of each building, it is necessary to first choose the PV technology that will be used. Because there are so many different solar panel technologies available on the market today, the panel module that was determined to be the most suitable for the study location was selected based on its ability to meet the requirements for widespread implementation. Mono NeoN LG280N1C 280 W, SW 280 Silver Mono Plus, SL280CE-36P) and MULTICRSTALLINE PM 280 are

Table 5.1 Table showing solar insolation values at sample locations

S. No.	Object ID	Solar insolation (kWh/m2/day)
1	2	7.439236
2	6	7.537281
3	7	4.067364
4	8	7.280376
5	10	5.226242
6	15	0.478394
7	17	0.006588
8	18	5.876764
9	19	0.775425

the key panels now topping the market, according to markets and research. These panels were chosen based on their Ontario distributors' recommendations for cost, performance, and availability. The available surface area for panel installation has to be evaluated to establish how many panels could fit into the desired rooftop of a structure.

Solar PV specifications		
1	Technology type	Multicrystalline
2	Manufacturer	Central Electronics Limited
3	Model	PM 280
4	Power max (P_{mpp})	280 W
5	Maximum system voltage	1000 V
6	Voltage at P_{mpp}	35.00 V
7	Open circuit voltage	44.00 V
8	Current at P_{mpp} (I_{mpp})	8.00 A
9	Short circuit current	8.80 A
10	Fuse rating	15A
11	Irradiance (at 25 °C)	1000 W/m^2
12	Field wiring copper	14AWG
13	Physical dimensions	982.05 mm (Width) × 1960.1 mm (Length)

(a) ***Area Calculations for PV Installations***

(1) *The area occupied by Single Panel:*

$$\text{Area} = \text{Length} \times \text{Breadth} = 982.05 \times 1960.1\ \text{mm}^2 = 1924916.205\ \text{mm}^2$$

Area of each panel = 1.924916 m^2–1.93 m^2

Table 5.2 Usable roof area and solar PV available (Scenario 1)

Building Name	Total Area (m^2)	Usable area (m^2) (Total area × 0.30875)	*Number of panels	PV array size (kW)
Admin-building	2400	741.0	~ 384	64.512
SES	60 × 60	1111.5	~ 576	96.768
SBS	60 × 60	1111.5	~ 576	96.768
SSBS	60 × 60	1111.5	~ 576	96.768

* *Considering the above-mentioned panel of size 1.93 m^2 with 168 Watts generation potential*

(2) *The number of Panel Required:*

Number of Panel (N_p) =Area Available for PV installation/Area of Each Panel
Number of Panel (N_p) = 100 m^2/1.93m^2 = 51.81–52 Panels
(Here we have considered an available area for installation as 100 m^2)

(3) *The Power generated by Complete Grid:*

Power Maximum (P_{mpp}) =280 Watts
Considering Generation efficiency as 60 %, Power Max. (Pmpp) will be 168 Watts
Power generated by each panel= 168 Watts
Total Number of Panels=52 Panels
Total Power generated by 52 Panels= 168 Watts × 52 =8736 Watts=8.736 kW

Scenario 1: *Useable Roof Area and Available Solar PV computed with GIS.*

Table 5.2 explains the solar power production capacity for the chosen building roofs of the various educational institutions, including the School of Business Studies (SBS), the School of Earth Sciences (SES), and the School of Social and Behavioural Sciences (SSBS) (School of Social and Behavioural Sciences).

Scenario 2: GIS determined usable roof area and solar PV available.

According to the technique stated in earlier Sect. 5.4.8, the rooftops were analysed for a second scenario using GIS to estimate the area appropriate for PV deployment as well as the PV array size and capacity. The module coverage factor of 0.475 was multiplied by the useful area (in metres squared) that was scanned by the GIS to determine the number of panels and the size of the PV array that would be required for each structure.

You can see the digitized photographs that were used to find the useable rooftop area in Sect.5.4.8 and Table 5.3 describes those photographs. The results for Scenario 2 are summarized in Table 5.3 which presents GIS-based usable areas together with the PV array sizes and capabilities for every structure in the study area. Table 5.2 explains the solar power production potential for the chosen building roofs of several schools, including the School of Earth Sciences (SES), the School of Business Studies (SBS), and the School of Social and Behavioural Sciences (SSBS) (School of Social and Behavioural Sciences).

Table 5.3 Usable roof area and solar PV available (Scenario 2)

Building name	Total area (m^2)	Usable area (m^2) (Total area × 0.475)	Number of panels	PV array size (kW)
Admin-building	2400	1140	~ 590	99.120
SES	60 × 60	1710	~ 886	148.848
SBS	60 × 60	1710	~ 886	148.848
SSBS	60 × 60	1710	~ 886	148.848

* *Considering the above-mentioned panel of size 1.93 m^2 with 168 Watts generation potential*

5.6 Summary

Incoming solar radiation from the sun, also known as insolation, is the primary source of energy for a great deal of the earth's physical and biological activities. For a comprehensive understanding of a broad variety of environmental and human systems, it is essential to have a solid grasp of their relevance to landscape sizes. With the solar radiation analysis tools of the ArcGIS spatial analyst extension, you can map and analyse the effects of the sun's radiation over a geographic region for a range of different periods. It takes into account the influences of the atmosphere, the latitude and elevation of the location, the slope and the direction of the compass (known as the aspect), the daily and seasonal shifts in the angle of the sun, and the shadows produced by the terrain of the surrounding area. The outputs that are produced therefore are easy to integrate with other GIS data and may be used in the modelling of physical and biological processes to examine how the sun affects them. When doing an examination of solar radiation for a landscape or a particular area, we employed two different approaches:

(a) Utilize the Area Solar Radiation tool to get an accurate reading of the insolation throughout an entire landscape. To produce insolation maps over a vast region, the computations are performed many times, one for each place on the topographic surface that was provided.
(b) The quantity of radiant radiation emitted by the sun at a certain area may be determined using the Points Solar Radiation tool. Locations may be recorded in a location table either as point features or as x, and y coordinates. Both options are available. Only at certain locations is it possible to do calculations of solar radiation.

The solar radiation graphics tool can be utilized to produce graphical illustrations of the visible sky (viewshed map), the sun's position in the sky over time (sun map), and the sectors of the sky that influence the amount of incoming solar radiation for analytical research (sky map). During the analysis process, these "maps" are used, in principle, to compute the total amount of solar radiation that is present in a certain place or region. This article demonstrates that both global and diffuse radiation knowledge are required to compute the hourly slope radiation and so make the best use of available solar energy (by selecting an optimal combination of tilt and orientation). The study can be extended to cities/sites where diffuse radiation is not measured, even if both global and diffuse radiation were provided for the currently used locations.

As a result of the lack of ready-to-use solar radiation data at the place of interest, it is necessary to estimate radiation. The most common manifestation of a solar application, the calculation of radiation received by inclined surfaces, requires both diffuse and direct components. This emphasizes the importance of estimating either of them from more widely observed global radiation to obtain the third component as the difference between the two. These devices come in a wide variety of shapes and sizes. The primary purpose of this study was to construct models that would yield more accurate yet computationally simple radiation datasets. Incorporating commonly available meteorological factors into classic estimation models can improve the approach even more. This study combines conventional and newly created urban archetypes to measure the influence of urban forms on energy consumption and energy system design. A multicarrier energy hub measures urban morphology's impact on energy consumption. Urban archetypes impact the heating and cooling energy consumption of city districts categorized by form characteristics and floor area ratio. Demand indicators can't quantify impact. Urban forms may increase energy system costs by up to 50% over peak and/or yearly demand. Urban structures affect the integration and usage of renewable energy. Beyond building modelling, urban planning should integrate energy system design.

References

Bergamasco L, Asinari P (2011) Scalable methodology for the photovoltaic solar energy potential assessment based on available roof surface area: application to piedmont region (Italy). Sol Energy 85(5):1041–1055. https://doi.org/10.1016/j.solener.2011.02.022

Buyantuyev A, Wu J (2010) Urban heat islands and landscape heterogeneity: Linking spatiotemporal variations in surface temperatures to land-cover and socioeconomic patterns. Landscape Ecol 25(1):17–33. https://doi.org/10.1007/s10980-009-9402-4

Chaudhari S, Felfelani F, Shin S, Pokhrel Y (2018) Climate and anthropogenic contributions to the desiccation of the second largest saline lake in the twentieth century. J Hydrol 560:342–353. https://doi.org/10.1016/j.jhydrol.2018.03.034

Gao Y, Chaudhari P (2021) A free-energy principle for representation learning. Mach Learn Sci Technol 2(4). https://doi.org/10.1088/2632-2153/abf984

Mbow C, Smith P, Skole D, Duguma L, Bustamante M (2014) Achieving mitigation and adaptation to climate change through sustainable agroforestry practices in Africa. Curr Opin Environ Sustain 6(1):8–14. https://doi.org/10.1016/j.cosust.2013.09.002

Pramanick S, Ahmad N, Jain DK (2021) Real time electricity market (RTM) in India. In: 5th International conference on smart grid and smart cities, ICSGSC 2021, pp 112–117. https://doi.org/10.1109/ICSGSC52434.2021.9490431

Rey-Mahía C, Álvarez-Rabanal FP, Sañudo-Fontaneda LA, Hidalgo-Tostado M, Suárez-Inclán AM (2022) An Experimental and numerical approach to multifunctional urban surfaces through blue roofs. Sustainability (Switzerland) 14(3). https://doi.org/10.3390/su14031815

Sicard P, Augustaitis A, Belyazid S, Calfapietra C, de Marco A, Fenn M, Bytnerowicz A, Grulke N, He S, Matyssek R, Serengil Y, Wieser G, Paoletti E (2016) Global topics and novel approaches in the study of air pollution, climate change and forest ecosystems. Environ Pollut 213:977–987. https://doi.org/10.1016/j.envpol.2016.01.075

Tarigan E, Djuwari, & Purba, L. (2014) Assessment of PV power generation for household in surabaya using solarGIS-pvplanner simulation. Energy Proc 47:85–93. https://doi.org/10.1016/j.egypro.2014.01.200

Wang Y, Ren Y, Song L, Xiang Y (2021) Responses of extreme high temperatures to urbanization in the Beijing–Tianjin–Hebei urban agglomeration in the context of a changing climate. Meteorol Appl 28(5). https://doi.org/10.1002/met.2024

Yu X, Guo X, Wu Z (2014) Land surface temperature retrieval from landsat 8 TIRS-comparison between radiative transfer equation-based method, split window algorithm and single channel method. Remote Sens 6(10):9829–9852. https://doi.org/10.3390/rs6109829

Chapter 6
Variability of Energy Resources at a Diverse Scale

Abstract Analysing and characterizing the swing in solar insolation across space and time require a technique known as spatio-temporal and hyper-temporal variability analysis. Multiple measurements of a quantity provide different values. Spatial variability. The range measures spatial variability. Identifying regional fluctuations in solar insolation and their impacts on energy production may be helpful in specialized energy management systems. It's the study of unpredictability in science. In this portion, we looked at the techniques a researcher uses to examine a topic, including variability and reasoning. The researcher must know research and analytical methodologies. Researchers must know how to undertake variability analysis and employ research techniques. They must also know which techniques or processes are significant, what they mean and why. Researchers must grasp the assumptions behind certain approaches and the criteria for determining whether certain techniques and procedures are acceptable.

Keywords *Spatio-temporal* · *Hyper-temporal* · *Variability* · *Solar insolation*

6.1 Introduction

Understanding the spatial distribution of solar radiation is becoming more significant because of the increased interest in solar energy consumption (Gambhir et al. 2010). It is essential to have accurate knowledge about solar resources for a wide range of technical areas, such as agriculture, meteorology, forestry engineering, and water resources, for ground-breaking technologies such as solar energy (Ajayi et al. 2016; Rn et al. 2014). A comprehensive understanding of solar resources is essential for increasing production and fostering a more widespread use of solar energy as a source of power (Tanskanen et al., 2021). At every given location on the earth's surface, incident solar radiation is produced due to the complex energy transfers that take place in the middle of the atmosphere and the land or ocean surfaces below (Kumar, 2019a; Wu et al. 2013; Xie et al. 2021). On a global scale, latitudinal radiation disparities are caused by the geometry of the planet's orbit around the sun as well as the planet's rotation. Changing atmospheric conditions, which are mostly created by predictable astronomical factors (via solar altitude), are a major contributor to the

D. Kumar, *Renewable Energy Scenarios in Future Indian Smart Cities*, Advances in Geographical and Environmental Sciences, https://doi.org/10.1007/978-981-19-8456-3_6

seasonal and daily shifts that occur (e.g., clouds, aerosols, water vapours, and ozone). There is a large variety of variances in both the location and the time at which solar radiation reaches the surface of the earth (Ahmad and Goparaju 2018; Bhakta and Mukherjee,2017; Borfecchia et al. 2014). This volatility has an effect not only on the ability of electrical systems to be predicted but also on their ability to be managed.

The accuracy of geographical or temporal sun irradiation information enables us to increase the dependability (lower risk) and quality of solar systems, which is reflected in the pricing of the energy produced (Ali et al. 2020; Islam 2018). Solar radiation obtained from the database of the National Centers for Environmental Prediction's (NCEP) and Climate Forecast System Reanalysis (CFSR) was applied in this study to examine the spatial and temporal variations of solar isolations within the Kalaburagi (Gulbarga) District between the years of 2004 and 2014. Daily data were interpolated using a technique called standard kriging to calculate the levels of solar radiation for sites where there was insufficient information.

6.2 Sub-objectives

1. An investigation of the spatial and temporal irregularity of solar insolation.
2. An investigation of the hyper-temporal inconsistency of solar insolation.

6.3 Workflow Overview

The next stage of work on the thesis is shown in Fig. 6.1, which shows the broad process of the methodologies that were utilized. The process flowchart shown in Fig. 6.1 consists of three basic components: geographical variability, temporal variability, and hyper-temporal variability (Busho et al., 2021; Kumar 2019b; Yang et al. 2020). Each of these components describes an activity that must be carried out to obtain the required analytical conclusions. In addition to the three primary steps that have been outlined above, it also includes tasks such as (1) acquiring and manipulating data, (2) aggregating incident global solar radiation data, (3) segregation of different data parameters such as precipitation, wind, maximum and minimum temperature values, and (4) presenting data to perform spatial, temporal, and hyper-temporal variability estimates.

The technique used to conduct the study is the first component of the future methodology. Following that, the study region was chosen, followed by the acquisition, processing, and assessment techniques for the required datasets. The essential datasets for the research region were also examined for compatibility and, if necessary, adjusted to allow for more extensive analysis with the chosen solar programme.

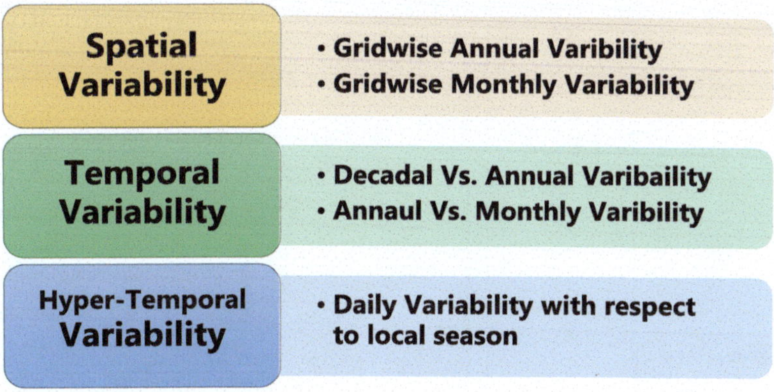

Fig. 6.1 Illustration to describe the complete workflow

6.4 Methods

The interannual stochastic fluctuation of solar radiation, which is vital for calculating the cost of the energy generated and directly impacts the enterprise's economic viability, has been made for the forecast of yearly energy output (Kestens et al. 2011; Saeed and Ahmad, 2021). In three ways, the solar resource is a requirement for the project of solar systems: (1) An analysis of the appropriateness of the location for a solar power plant; (2) A projection of the annual, monthly, or daily total energy produced; and (3) Solar irradiation forecasts and operational approaches.

The topography is a fundamental component that controls the spatial variability of insolation at landscape sizes (Islam 2018; Zhao and Fong, 2017). Several factors influence the amount of sunlight that reaches different locations, including elevation, direction (including slope and aspect), and the shadows cast by topographic features (Banzhaf et al. 2021; Ramachandra et al. 2012; Snir et al. 2016). This fluctuation varies with the time of day and the year, contributing to the variability of microclimate. Microclimate consists of characteristics such as air and soil temperature regimes, evapotranspiration, snowmelt patterns, soil moisture, and photosynthetic light availability (Antoszewski et al. 2022; Ellis et al. 2000; Guan 2011). This fluctuation varies with the time of day and the year.

6.4.1 Research Area

The analysis was done for the Kalaburagi (formerly Gulbarga) District in northern Karnataka, which covers 10,951 km^2. The district's average altitude is 300–750 metres. Krishna and Bhima are major rivers. This area has three seasons. Summer temperatures range from 38 to 44 degrees Celsius from late February to mid-June. From late June to late September, the southwest monsoon brings 750 mm (27–37 °C)

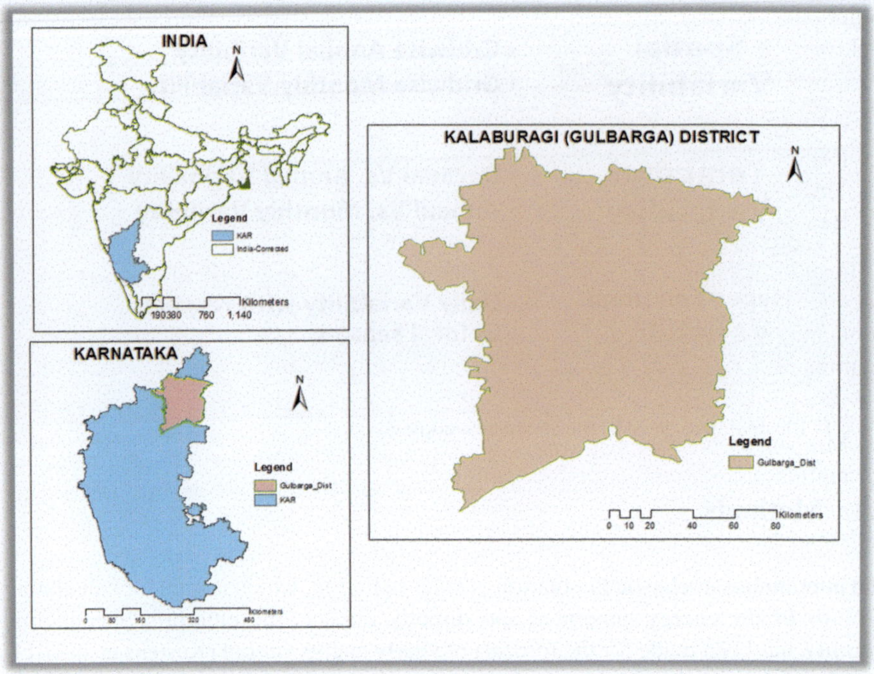

Fig. 6.2 Location of study area

of rain. Dry weather (11–26 °C) follows until mid-January. Figure 6.2 shows the study region.

Why the Kalaburagi as a research location?

To broaden the scope of the investigation, an experimental district was chosen and the reasons for opting for Kalaburagi as a research site have more general reasons:

1. It has a self-controlled territory, there are no vague boundaries.
2. Landscape character is unique and very varied, with various biological and landscape zones.
3. The researcher is also very much familiar with the location.

6.4.2 Datasets Used

This research uses handheld GPS, Survey of India 1:50 K topographic sheet, SRTM DEM/BHUVAN DEM, and Google Earth VHR satellite data (site validations). Other free archive data sets for execution and validation from BHUVAN NOEDA (NRSC), Landsat Archive and Earth Explorer (USGS). The following parameters were used to get data on solar radiation and other meteorological factors from multiple open archives:

1. There are three different kinds of solar irradiance: global solar irradiance, direct normal incidence (DNI) solar irradiance, and diffuse solar irradiance.
2. Rainfall and atmospheric pressure.
3. The direction and velocity of the wind.
4. Relative humidity and ambient air temperature.

In addition to in-situ data, information was acquired from a variety of open-source or educational websites. Additional information was gathered from books, journals, publications, and websites. The following are some of the many information sources mentioned:

1. Online gateways and databases *(ResearchGate).*
2. Ministry of New and Renewable Energy *(Website, Reports, Manuals)*.
3. Online Journals *(e.g., Solar Energy, Energy, Journal of Solar Energy Engineering).*
4. Books in the context of infrastructure management, trends in energy sciences.
5. Reference texts from ESRI, ERDAS Inc. publications, and manuals.

6.4.3 Software and Tools Used

1. Energy Plus 8.0
2. Geospatial Toolkit *(NREL)*
3. RET Screen Plus Software
4. ESRI ArcGIS 10. x (*for vector data analysis*).

6.4.4 Spatio-Temporal (S-T) Variability Analysis

Traditional spatio-temporal analysis has been rethought unexpectedly. The study was designed to be carried out on both a geographical and temporal scale. The examination of variability was split into two parts. The first part of the investigation focused on the spatial variability of solar insolation. Using datasets from the year 2013, the geographic variability was calculated on an annual basis for each of the grid cells. The solar insolation variability for each grid cell was tabulated. Later, we calculated the variability of solar resources for each grid cell month by month for the year 2013, in other words, we computed the variability of solar resources for each grid cell month by month for the year 2013. Every month, the annual mean (i.e., a yearly average of monthly values of each site) of all grid cells was used to represent the fluctuation of solar insolation. In a nutshell, the stated spatio-temporal variability analysis was expressed using the term "temporal variability on the spatial scale," which was developed recently. The assertion that a temporal variability analysis was done for each of the geographical locations is revealed. The obtained data were plotted and shown using standard standards for easy comprehension.

Finally, using typical "directed analyses," the geographical analysis sought to map out the variability. These approaches assisted us in recognizing the differences in solar insolation values that may be found in each of the eight cardinal directions, which are as follows: east, west, south, north, northeast, southeast, southwest, and northwest. The findings of this investigation were collected and analysed to establish whether or not there was a significant amount of variability in a certain path.

The second portion of the research looked at the temporal fluctuation of solar insolation quantities. On a time, scale, the horizontal and vertical fluctuation of solar insolation was accounted for in this section of the analysis. The analysis was carried out using two different methodologies. The first method included calculating the yearly variations of solar insolation at the decadal level. The yearly averages of each year were compared to the mean of decadal isolations for variability. The highest and minimum variability years were presented to determine which years of the decade were stable and which were unstable. The year with the most departures from the decadal mean was labelled an unstable year, while the year with the least differences was labelled a stable year for solar variability. The second technique involves estimating the yearly shift in solar insolation for exceptionally steady years as well as years that were extremely unstable over the decade (as determined by previous studies). The variability of solar resources was estimated for the indicated years every month in this section of the analysis, and the findings were provided as tables and maps. To express the findings of the variability study, suitable tables and charts were created.

1. ***Spatial Variability***

When we measure a quantity at multiple spatial locations, the values change across the sites, which is known as spatial variability. Geostatistics and geographical descriptive statistics can be used to examine spatial variability. The yearly solar insolation readings for each n × n matrix cell are compared to the surrounding cells in the matrix according to the pattern illustrated in Fig. 6.3, which defines the variability of the solar resource within the matrix. Each of the pixel values in the layout was subjected to the procedure. Variability statistics were computed using similar procedures on a yearly and monthly basis. These obtained variability statistics were utilized to plot the variability maps for a better visual representation.

The alignment or sketch depicted in Fig. 6.3 is used to compute the spatial variability within the research region. The whole research region was gridded according to the aforementioned arrangement, and the solar insolation values of each cell were used to calculate the variability for the next neighbouring cell. We can perceive from Fig. 3 that cell numbers **P1, Q1, R1, S1, and T1** have corresponding solar insolation values of $\mathbf{T_{11}}$, $\mathbf{T_{12}}$, $\mathbf{T_{13}}$, $\mathbf{T_{14}}$, and $\mathbf{T_{15}}$, respectively. Likewise, the diagonal cell numbers **P1, Q2, R3, S4, and T5** indicate the corresponding solar insolation values of $\mathbf{T_{11}}$, $\mathbf{T_{22}}$, $\mathbf{T_{33}}$, $\mathbf{T_{44}}$, and $\mathbf{T_{55}}$. Beyond these, the solar insolation values for other cell numbers **P2, Q2, R2, S2, T2…P5, Q5, R5, S5,** and **T5** were used identically to carry out the variability analysis.

Fig. 6.3 Framework for spatial variation analysis

	P	Q	R	S	T
1	T_{11}	T_{12}	T_{13}	T_{14}	T_{15}
2	T_{21}	T_{22}	T_{23}	T_{24}	T_{25}
3	T_{31}	T_{32}	T_{33}	T_{34}	T_{35}
4	T_{41}	T_{42}	T_{43}	T_{44}	T_{45}
5	T_{51}	T_{52}	T_{53}	T_{54}	T_{55}

The variability scheme framed in Fig. 6.3 identifies the variability for any cell number value from the mean of other cell values. For example, if we want of computing the variability of cell numbers $\mathbf{T_{11}}$, $\mathbf{T_{22}}$, $\mathbf{T_{33}}$, $\mathbf{T_{44}}$, and $\mathbf{T_{55}}$, respectively, then we will compute the variation of these cell number values from the mean of a complete set of cell values. To calculate the geographical variability, the same procedure was used to compute the means on a monthly (e.g., the means of all Januarys, Februarys, and so on) or yearly (i.e., the annual mean of the year) level, as shown in Fig. 6.4.

The actual setup for performing the spatial variability analysis in the research region is shown in Fig. 6.4. For the sake of accuracy and identity, the entire region has been gridded with several tiny cells, each of which is represented by a unique numerical code. These distinct numbers also indicate the levels of solar insolation, which were later utilized in the research.

Here, the standard deviations of the surrounding cells were calculated as

$$\sigma_s = \left[\sum_{i=1}^{n} (\overline{X} - X_i)^2 \times \frac{1}{n} \right]^{1/2} \tag{6.1}$$

And the spatial COV is

$$C_s = \sigma_s / \overline{X} \tag{6.2}$$

Solar insolation variability maps were created after the foregoing methods were completed to visualize the level of spatial variation happening in a year as well as the monthly basis for the reported area. After you've finished with the spatial variability on spatial extends, you may go on to the next step.

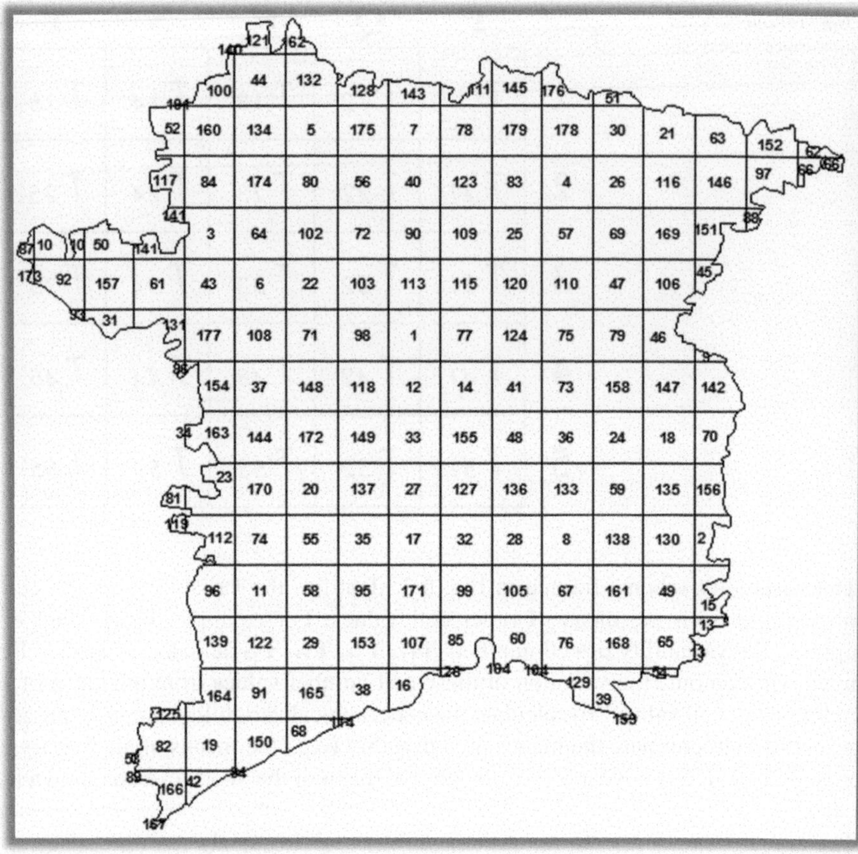

Fig. 6.4 Actual layout for spatial variability analysis

(a) *Spatial Variability of Annual Global Solar Insolation*

To compute and visualize the spatial variation, the yearly solar insolation values of each grid cell (as shown in Fig. 6.4 and mentioned at the beginning of Sect. 6.4.4 of this chapter) were used. The records for the year 2013 were used to compute the solar insolation variability in this study.

Figure 6.5 shows the easiest approach to depict the variability for each of the grids. However, due to dilutions or compactness in the data sets being displayed, we are frequently unable to obtain any useful information about the datasets. All of the significant peaks and dips were aggregated with excellent compactness, allowing us to find all of the answers to our study queries. As a result, following the variability analysis, to provide more relevant datasets. The current study added to this background by computing the geographical variability of solar insolation for each grid cell using the schematics illustrated in Fig. 6.4. (shown as grid cell numbers like 1,

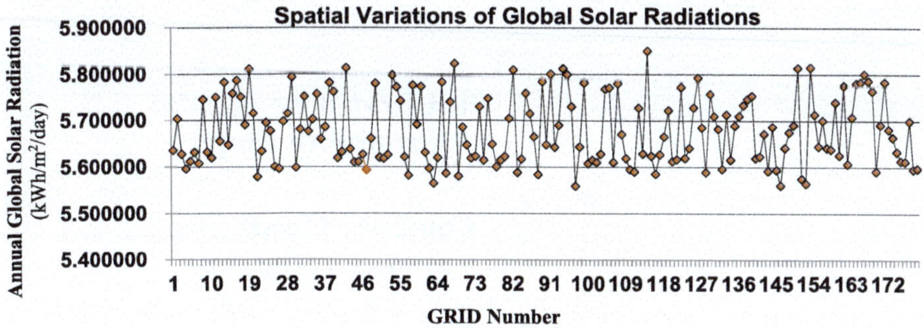

Fig. 6.5 Illustration for annual global solar radiation variability

2, 3, 4, 5, etc., in Fig. 6.4). The variability for each grid cell number, such as 1, 2, 3, 4, 5, 7, 9, and so on, is calculated by taking the mean of all grid cell solar insolation readings. The associated standard deviations and spatial covariance (SCOV) values were calculated using the expressions provided in Equations (6.1) and (6.2). After computing the values of each grid cell variation, the associated results are plotted and displayed in a tabular and graphical format for easier understanding.

(b) *Spatial Variability of Monthly Global Solar Insolation*

Similarly, for each of the grid cells, the values of spatial variabilities were computed every month for the year 2013 using the approach described in the previous sections. The monthly mean values of all grid cells were calculated, and the variability of each grid cell value was estimated. The geographic variation was calculated and shown with the use of the monthly solar insolation data for 2013. (as shown in the schematics of Figs. 6.3 and 6.4). The intermediate findings, as well as the outcomes, were collated and displayed to provide a clear picture of the riddles surrounding the microlevel process of variability. The associated standard deviations and spatial covariance (SCOV) values were calculated using the expressions provided in Equations (6.1) and (6.2). Every month for the whole year of 2013, the processes were repeated to compute the variability.

The technique for doing the variability analysis every month is illustrated in Fig. 6.6. This diagram depicts a self-explanatory pyramid, with the yearly mean of solar insolation numbers at the bottom and the produced variability plots and tables at the summit of the pyramid. The current phase of the study adjusted for monthly spatial solar insolation variations and reported the results for the study region on a temporal scale (month-by-month).

(c) *Directional Global Solar Insolation Spatial Variability*

In addition to the methods of variability analysis that have been mentioned, an effort was made to determine the amount of variability in each cardinal direction. These directions are as follows: east–west, north–south, northeast-southwest, and southeast-northwest. Each direction's variability was shown along its corresponding direction, and the directions with the least and largest variability were reported.

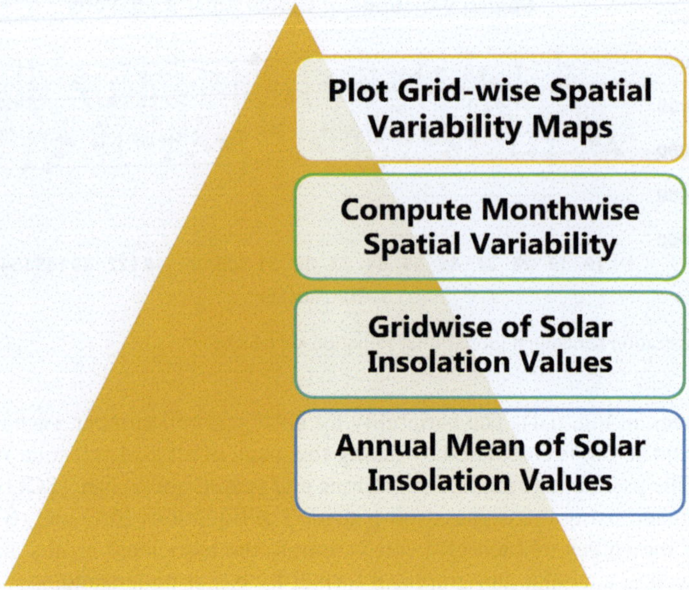

Fig. 6.6 Schematics for monthly spatial variability analysis every month

The setup for the methodologies used for directional variability analysis is shown in Fig. 6.7. The entire research area was partitioned into numerous grid cells, as shown in the imagined picture, according to the already described schematics (of Fig. 6.4), with grid cell number 13 at the grid matrix's centre. The cell in the centre of the grid, number 13, was used as a point of reference, as was the intersection of all of the directed lines (i.e., east–west, north–south, northeast-southwest, and southeast-northwest). All grid cell numbers that fell in line with directed lines were tagged (as illustrated in the schematics in Fig. 6.7), and solar insolation figures for all grid cell numbers were retrieved. For a better comprehension of the variability in the directions, the values obtained in the previous phases were recast according to the schematic of directions. Following that, the mean for each direction was calculated using the retrieved values and the procedures and methodologies previously stated. The relevant standard deviations and spatial covariance (SCOV) values were calculated using the equations specified in Equations (6.1) and (6.2). For each of the directions, the results were reported in tabular form.

2. ***Temporal Variability***

Temporal variability refers to the variations that occur over time and are also referred to as vertical and horizontal variations. The fluctuation of solar insolation on various timescales is generally explained by temporal variability. We used two ways to calculate the temporal variability of solar insolation values in this study. The first method based the decadal mean on yearly mean values to estimate solar insolation variability. It also determined which years had the most and least fluctuation. The second

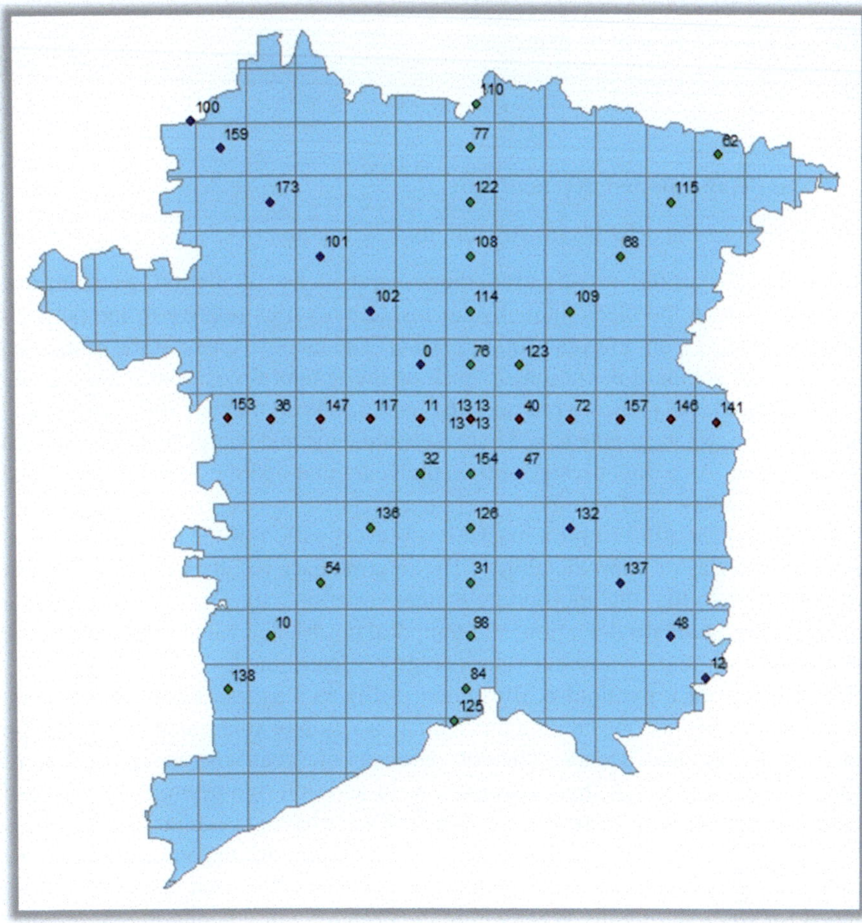

Fig. 6.7 Layout for directional spatial variability analysis

approach was more thorough because it compared the yearly mean solar insolation for years with high and low variability (as established by the first method) every month. Afterwards, the months with high variability, months with low variability, and months with anomalous variability were shown in the study that linked to monthly variability. Using estimations of the typical standard deviation and a temporal coefficient of variation (COV), we were able to measure the variability that occurs across time steps for each cell. In the present study, ten-year mean irradiance $\overline{X}$ were used to compute variation for each annual value X_i to derive the standard deviation of the data set.

$$\sigma_t = \left[\sum_{i=1}^{n} (\overline{X} - X_i)^2 \times \frac{1}{n} \right]^{1/2} \tag{6.3}$$

where $\overline{X} = \frac{1}{n}(X_1 + X_2 + X_3 + \ldots X_n)$

The temporal COV will be expressed as:

$$C_t = \sigma_t / \overline{X} \tag{6.4}$$

The COV is being reported as percentages.

(a) *Variations in Decadal Versus Annual Solar Insolation*

In the present part of the investigation, the researchers do calculations to determine the extent to which the mean annual solar insolation varies relative to the period's decadal mean (i.e., from the year 2004 to 2014). The inconsistency of the computed values with the standard deviations as well as the temporal covariance by making use of the formulas is presented in Sects 6.3 and 6.4. With the use of the decadal vs annual variance, we were able to see how the annual mean deviates from the decadal mean over time. As a direct result of this, the years that had the least, greatest, and most uncommon levels of variability were uncovered.

The chosen year allowed for a macro-level investigation of the reasoning underpinning the puzzles of unpredictability. The intermediary results were described in a tabular layout so that the appropriate annual variation values could be computed using the mean, standard deviation, and temporal covariance values. This was done so that the proper yearly variation values could be determined.

The structure of a pyramid is illustrated in Fig. 6.8 as the design that will be used for calculating and visualizing the annual variation of solar insolation from the mean over the previous decade. The computations of "year-wise variability" were the most important part of this investigation. In addition, the desired years within the decade were determined in this stage (which were having peculiar or anomalous trends). The decade's mean value serves as the foundation for the entire analysis. As a result, great care was taken while computing the mean of the solar insolation values included for study during the year (within the decade). For a better comprehension of the intricate ideas in the datasets created, the findings were presented in the form of tables, charts, and maps.

(b) *Variations in Annual Versus Monthly Solar Insolation*

The present section of the study estimates variability in the monthly mean solar insolation with the corresponding annual mean values for that particular year.

To complete the current study, the years with the highest, lowest (or maximum and minimum), and anomalous variability within the decade (i.e., from 2004 to 2014) in terms of their decadal mean values were chosen. These findings came from the analysis carried out in the previous section (i.e., a section dealing with the analysis of the annual mean concerning the decadal mean of solar insolation values). The calculation of the monthly mean values included rearrangement of these annual sets of statistics monthly. As was noted in the section that came before this one in the study, a similar kind of calculation was carried out to assess the standard deviations,

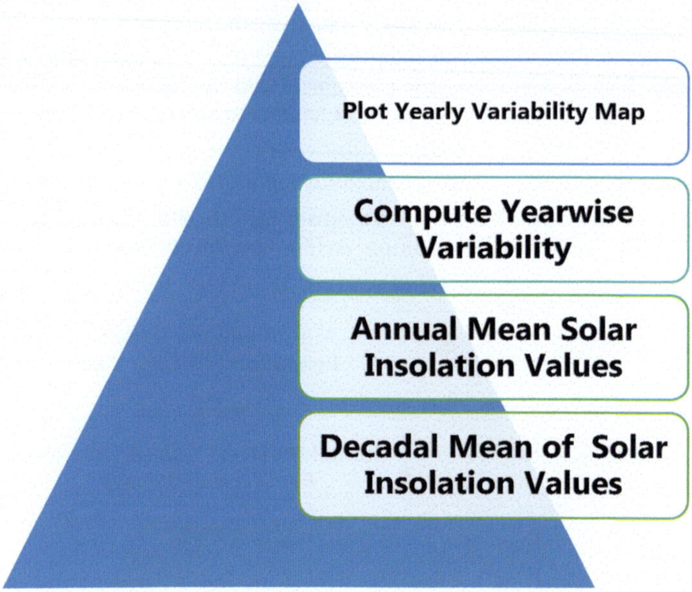

Fig. 6.8 Schematics for analysing the fluctuation of solar insolation on a decadal versus an annual timescale

temporal covariance values, and variability of solar insolation. The expressions that were used were presented in Equations (6.3) and (6.4).

The schematics provided in Fig. 6.9 demonstrate the whole sequence of all the processes necessary to accomplish this study. The pyramid confirms that the accuracy of the yearly mean of solar insolation numbers is critical to the entire study. The calculations of month-wise variability' were the most important part of this study. This step calculated the monthly mean value variabilities about the year's annual mean values. The same procedure was followed for a collection of values from various identifiable or atypical years. To guarantee that everyone was aware of the facts and data uncovered by the analysis, the results were presented in the most favourable daylight possible.

6.4.5 *Hyper-Temporal (H-T) Variability Analysis*

Because of the tremendous public interest in and responses to the use of solar photovoltaic potentials for diverse uses. It is now critical to examine the inconsistency in solar insolation on the most microlevel scale. In this case, a well-known geospatial technology approach may be expanded to do the needed analysis. With the given data sets, the approach may be customized for the present study and a new framework can be built for the analysis. The "*Hyper-Temporal Analysis*" approach may be used to

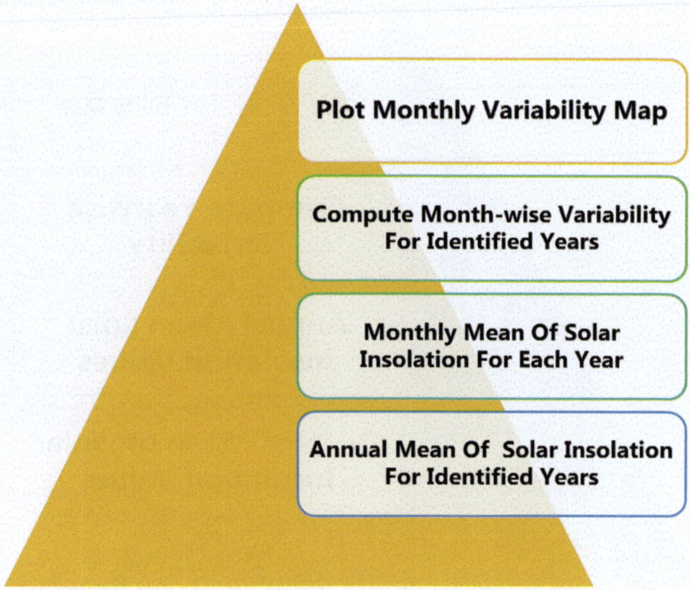

Fig. 6.9 Comparison of annual and monthly variations in solar insolation, with accompanying diagrams

analyse the variability of any parameter because it allows us to analyse our datasets at the lowest possible temporal scale.

In principle, hyper-temporal variability analysis is concerned together with the variation of any entity on a microtemporal scale, such as on an hourly, daily, monthly, or yearly basis. The possibilities and insights that may be gained by using hyper-temporal techniques have not yet been fully realized in their entirety. The currently available methods for identifying fluctuations in solar insolation using satellite data sets were found to be insufficient. These datasets have not yet been processed using methods that might be advantageous to the data treatment process. The occurrence and velocity of change in ratio to the minute interval frequency and geographical resolution of the data that are obtained will determine our ability to accurately monitor change on the precipice of the future. There is a possibility that complex problems, such as those requiring the evaluation of variability on a microlevel, may be solved by making use of hyper-temporal data in conjunction with more sophisticated analytic techniques. These included the development and validation of high-temporal (summer, monsoon, and winter) resolution tools to analyse the variability included in the solar insolation datasets that were readily accessible. To examine the temporal variation that exists within the study area due to the local seasons, the above methodologies are used. Table 6.1 is an illustration of the local seasons that occur in the Kalaburagi area.

According to Table 6.1, three major local seasons occur in the study area, and each one lasts for a different number of months. The months of February through June

Table 6.1 Local seasons in the Kalaburagi District

Summer-season	Monsoon-season	Winter-season
February		October
March	July	November
April	August	December
May	September	January
June		

make up the summer season, while July through September makes up the monsoon season, and the months of October through January make up the winter season.

The researchers attempted to assess the microlevel variability of solar insolations for the years 2010 and 2013 by using a method that was recently created and given the name hyper-temporal variability analysis. The year in question was selected for examination because it included a considerable number of observed anomalies (also known as increased monthly swings). This is one of the reasons why the year was chosen. The inquiry was carried out following the local seasons, with significant months that displayed distinctive characteristics being selected for further in-depth examination within the context of the local seasons. The investigation uncovered unexpected patterns, which provided researchers with a greater understanding of the unpredictability of trends as well as the mysteries that are associated with them.

The "hyper-temporal" analysis is shown pictorially in Fig. 6.10 via a series of schematics. A further implication of the pyramid structure is that the accuracy of the hyper-temporal inconsistency in solar insolation is affected, to some degree, by the month that is selected from the local season. The most crucial aspect of this inquiry was determining "a month within the local season" to transfer the variability estimates. This was the most important component. This stage monitored and computed the daily variations for the given set of mean values on monthly and daily basis while considering the local seasons. For the second set of monthly data compiled from the different local seasons, a technique quite like the first was followed. Because we wanted to guarantee that everyone understood the research, the final findings were presented in a way that included graphics.

6.5 Results and Discussions

To comprehend the variations in solar energy, it is necessary to first construct a framework that considers both time and space. Agriculture, forestry, ecology, and biology are some of the fields that may make use of such methods in situations when the regional fluctuation of solar radiation is more significant. To analyse the spatio-temporal and hyper-temporal data for variation mapping, several different approaches for variability analysis were performed. These methodologies gave a variety of evidence about the differences. The technique was used in monthly data bins to investigate the fluctuations from a seasonal point of view (e.g., the eighth of

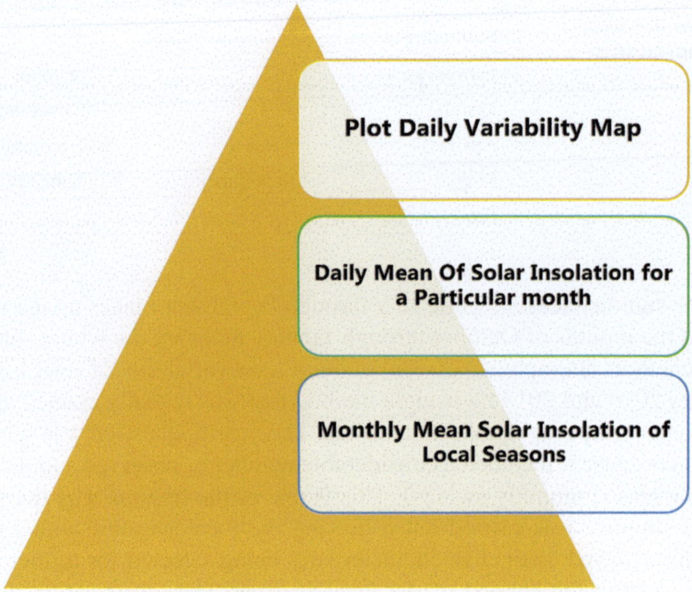

Fig. 6.10 Representations for hyper-temporal solar insolation variability analysis

January, February, etc.). A measure of the variability of solar resources through time and place was represented using tables, figures, graphs, and maps at the appropriate output scale. This was done to convey the information. The findings of this inquiry into the spatio-temporal and hyper-temporal variability of data about solar insolation were rather fascinating, to say the least. The following is a list of the primary outputs that may be obtained by using each analytical method:

6.5.1 Outcomes from Spatio-Temporal Variability Analysis

1. ***Spatial Variability Results***

(a) *Variability of Annual Spatial Global Solar Insolation*

The disparity between how solar insolation is distributed throughout the landscape and how it should be is brought into sharp relief by spatial variability. The recently established method of variability analysis has the potential to exploit this disparity more effectively. As was said before, the method that we were using at the time enabled us to recognize patterns of variability that were present in the area that was being investigated. To conduct the variability analysis, the present research made use of solar insolation data collected in the year 2013.

Figure11 shows that the top part (i.e., the northern half of the district) has the most variability, indicating that the source of solar insolation is unstable, whereas the bottom part (i.e., the southern part of the district) has the least fluctuation, indicating that the source of solar insolation is stable. This is demonstrated by the fact that the bottom part has the least fluctuation. From these two pieces of information, we can conclude that the area has a diverse collection of solar resources. Furthermore, we know that areas with low variability are optimal for the installation of photovoltaic power plants, whereas areas with high fluctuation are not ideal for the installation of any photovoltaic project because of the irregularity and volatility of the radiation coming from the sun.

(b) *Spatial Variability of Monthly Global Solar Insolation*

Every month during the whole year of 2013, the spatial variability of global solar insolation was displayed (as illustrated in Fig. 6.12). Individual spatial variability maps give a synoptic picture of monthly solar insolation fluctuations for each grid

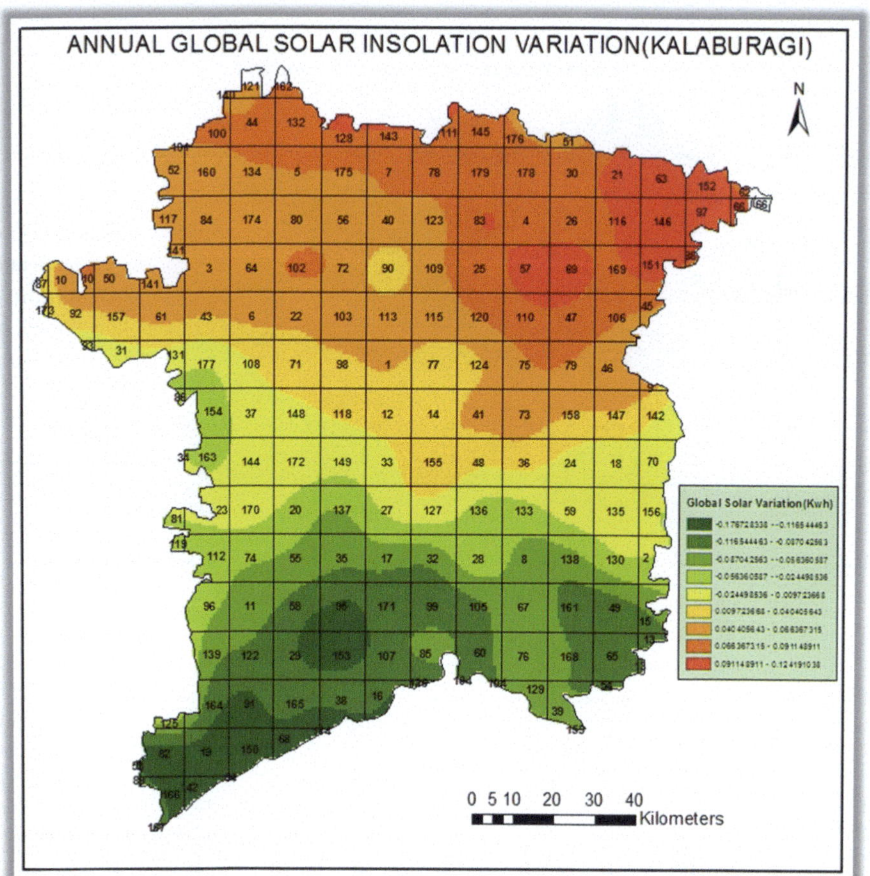

Fig. 6.11 Annual global solar radiation variability map

cell. Monthly spatial variability analysis was carried out using techniques comparable to those used in yearly spatial variability analyses. Only in the yearly and monthly contexts is it delayed. The annual mean (i.e., yearly average of monthly values of each location) of all grid cells was used to calculate the variability for each of the grid cells each month. The whole chain of variability for each month of that specific year was computed using the same techniques. We were able to determine the months with the lowest, highest, and most unusual variability as a result of our investigation.

To visualize the notion, the measured levels of variability for each month were shown against matching month-wise variability maps. These are shown in Fig. 6.12a through k, respectively.

The specifics of the solar insolation variability map from January to December are shown in Fig. 6.12 a and k. These maps were created by calculating the monthly variations from the yearly mean of solar insolation. It is quite easy to see at each GRID level how much the amount of solar insolation varies greatly from neighbouring GRID sites. In other words, we can see how solar insolation has changed every

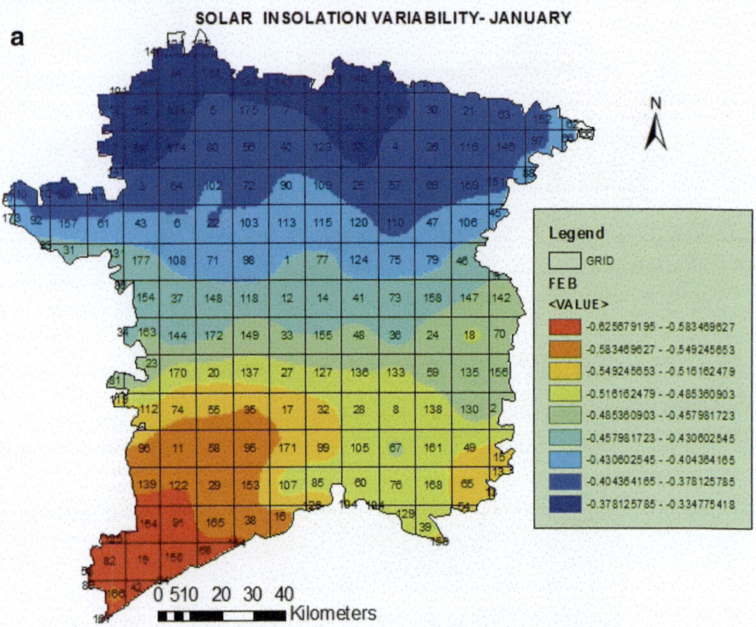

Fig. 6.12 **a** Monthly global solar radiation variability map for January month, **b** Monthly global solar radiation variability map for February month, **c** Monthly global solar radiation variability map for March month, **d** Monthly global solar radiation variability map for April month, **e** Monthly global solar radiation variability map for May month, **f** Monthly global solar radiation variability map for June month, **g** Monthly global solar radiation variability map for July month, **h** Monthly global solar radiation variability map for August month, **i** Monthly global solar radiation variability map for September month, **j** Monthly global solar radiation variability map for October month, **k**Monthly global solar radiation variability map for November month, **l** Monthly global solar radiation variability map for December month

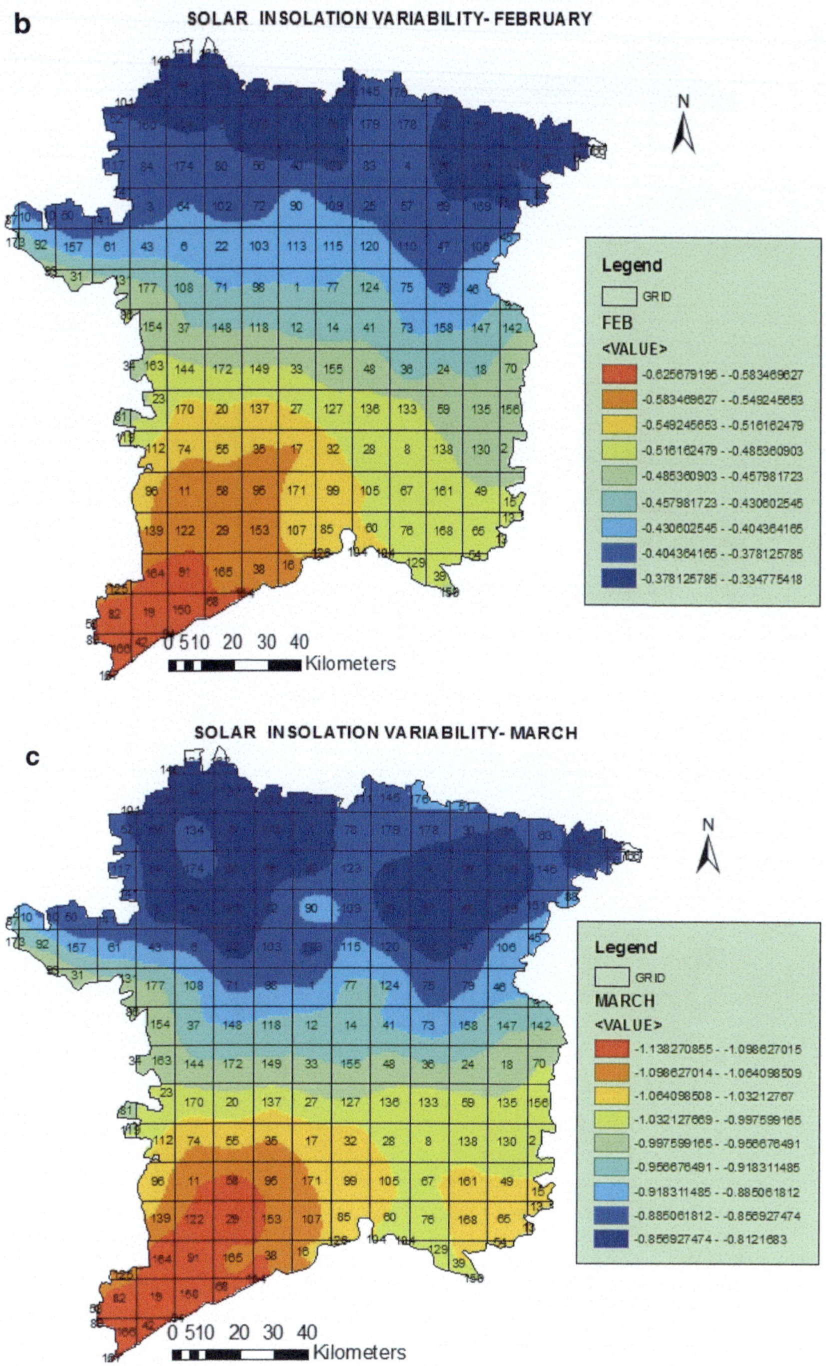

Fig. 6.12 (continued)

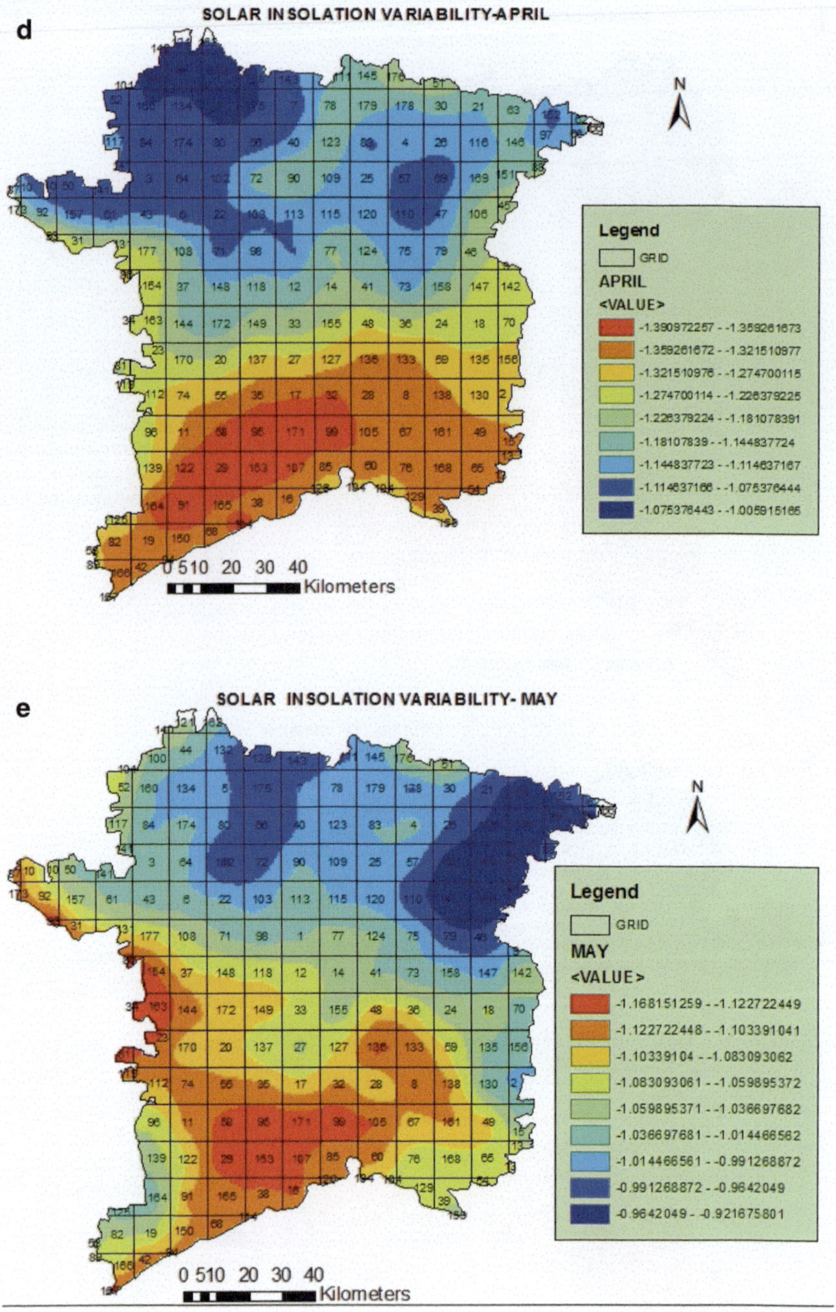

Fig. 6.12 (continued)

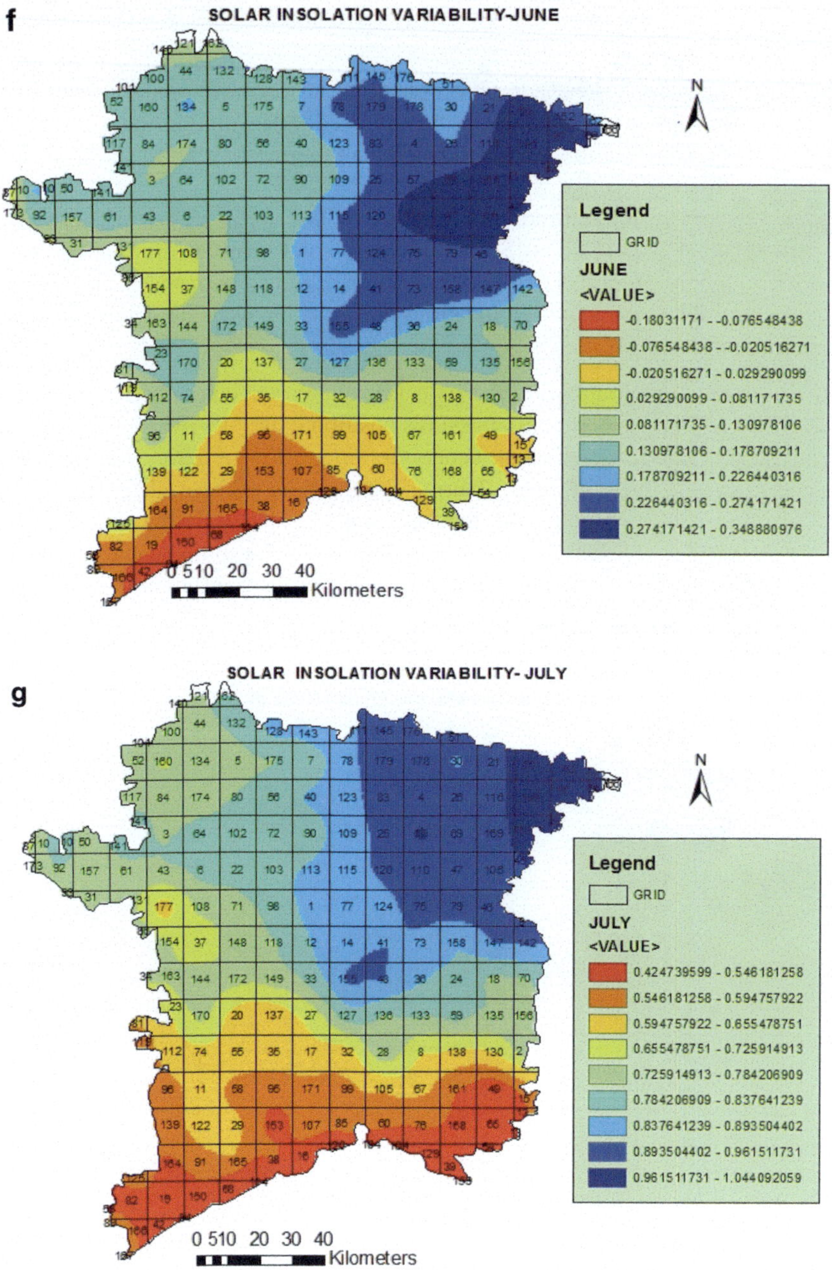

Fig. 6.12 (continued)

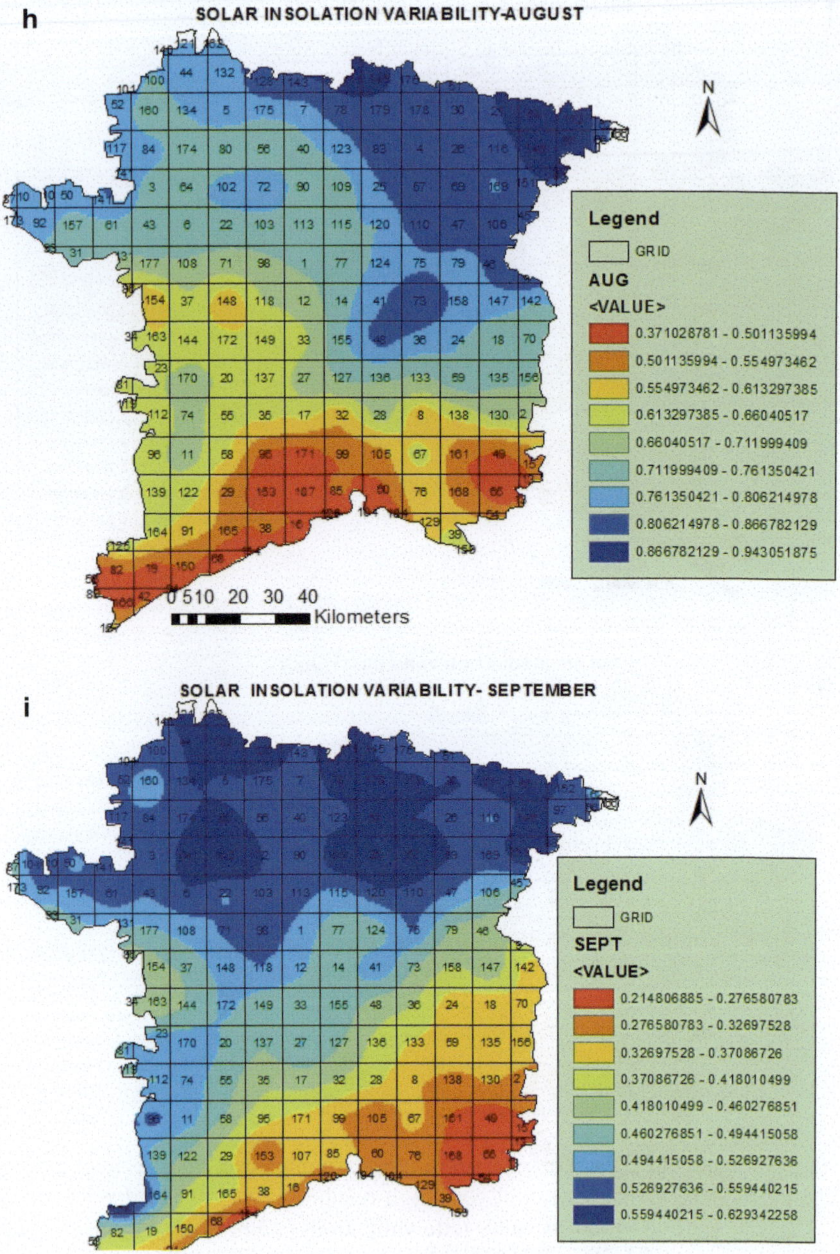

Fig. 6.12 (continued)

j

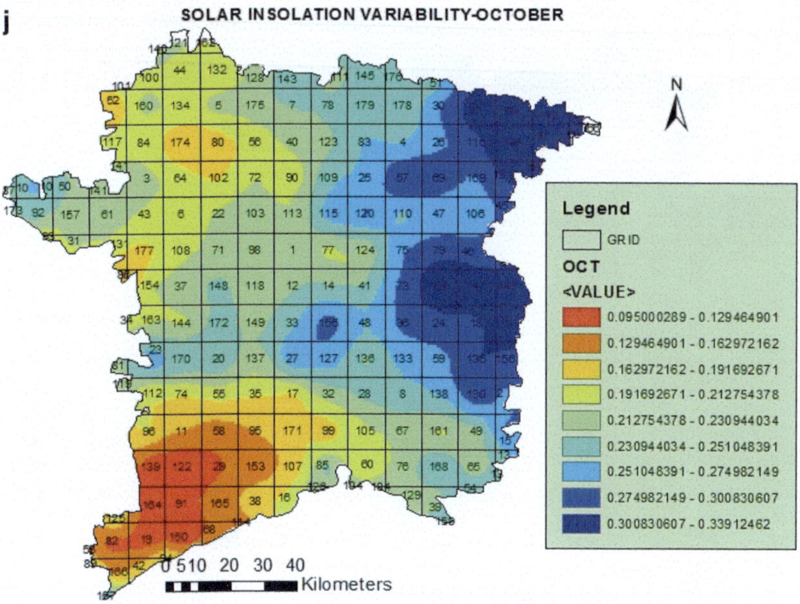

k

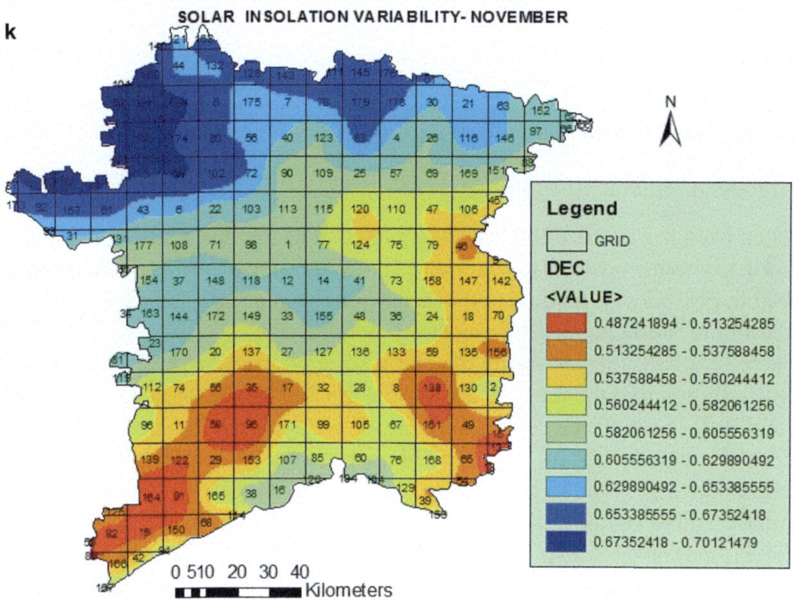

Fig. 6.12 (continued)

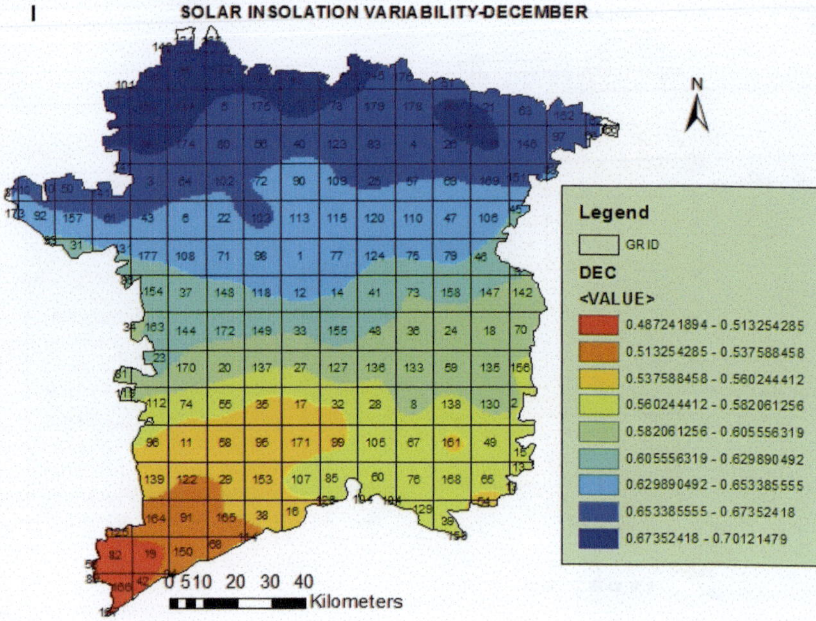

Fig. 6.12 (continued)

month throughout the year. Every month, these variability maps show a synoptic picture of the variance within the Kalaburagi district.

(c) *Directional Spatial Variability*

It has been tried to visualize the variability in each of the directions, namely east to west, north to south, northeast to southwest, and southeast to northwest. The approach described in Fig. 6.7 was applied to datasets from the year 2013, and the results were tabulated in the following order for each of the directions.

Table 6.2 explains the variability of solar insolation in an east–west direction. The said table exhibits that there exists a very meagre amount of variability. The values of variability for each of the cells falling in line with the said direction were used to compute the mean variability in a particular direction. The mean variability of the solar insolation was found to be 0.019781 kWh/m^2/day.

Similarly, Table 6.3 describes the variability of the solar insolation in a north–south direction. The table also summarizes the values of global solar insolation values concerning spatial locations. The amount of variability in global solar insolation values has been also reported in the table. From the values recorded in the table, we can formulate the understanding that there is a negligible amount of variability in the said direction. The mean of global solar insolation values was found to be 0.00461285 kWh/m^2/day. This value is so small that, it will not cause any effect on the variability of the solar insolation in the said direction.

Table 6.2 Variability in east west direction

FID	Lat	Long	GHI (kWh)	Var. GHI
11	76.85	17.05	5.65517726	0.023464
13	76.95	17.05	5.646905479	0.031736
36	76.55	17.05	5.685855342	-0.00721
40	77.05	17.05	5.632490959	0.04615
72	77.15	17.05	5.622513699	0.056128
117	76.75	17.05	5.66518411	0.013457
141	77.442	17.0461	5.671424932	0.007217
146	77.35	17.05	5.641055342	0.037586
147	76.65	17.05	5.674902192	0.003739
153	76.462	17.0529	5.713855342	-0.03521
157	77.25	17.05	5.63810411	0.040537
Mean			**5.658860797**	**0.019781**

Table 6.3 Variability in north-south direction

FID	Lat	Long	GHI(kWh)	Var. GHI
13	76.95	17.05	5.6469055	0.03173597
31	76.95	16.75	5.7521318	-0.0734903
76	76.95	17.15	5.6488266	0.02981487
77	76.95	17.55	5.6004762	0.07816529
84	76.943	16.5543	5.7593419	-0.0807005
98	76.95	16.65	5.7816373	-0.1029958
108	76.95	17.35	5.6184608	0.06018063
110	76.9641	17.631	5.5900104	0.08863104
114	76.95	17.25	5.6236425	0.05499898
122	76.95	17.45	5.6189479	0.0596935
125	76.9189	16.4937	5.792654	-0.1140125
126	76.95	16.85	5.684651	-0.0060095
154	76.95	16.95	5.644686	0.03395542
Mean			**5.6740286**	**0.00461285**

Table 6.4 summarizes the amount of variability in the northeast-southwest direction. Here, we can observe that a slight amount of variability exists. The table also explains the amount of solar insolation available at each of the reported locations. The account of variability in solar insolation for each of the locations was reported for the said direction. It was observed that there is a variability of 0.017817539 kWh/m^2/day exists in the said direction.

Table 6.4 Variability in northeast-southwest direction

FID	Lat	Long	GHI(kWh)	Var. GHI
10	76.55	16.65	5.750626301	0.071984852
13	76.95	17.05	5.646905479	0.03173597
32	76.85	16.95	5.676493973	0.002147477
54	76.65	16.75	5.742786575	0.064145126
62	77.4466	17.5379	5.564631233	0.114010217
68	77.25	17.35	5.580136986	0.098504463
109	77.15	17.25	5.594612877	0.084028573
115	77.35	17.45	5.585095616	0.093545833
123	77.05	17.15	5.640281644	0.038359806
136	76.75	16.85	5.733870959	0.055229509
138	76.4617	16.5523	5.75362137	0.07497992
Mean			**5.66082391**	**0.017817539**

Likewise, Table 6.5 reports the variability of solar insolation in the southwest-northwest direction. It was observed that the very least amount of variability with the value of 0.005446495 kWh/m^2/day exists in the said directions.

If we compare the mean variability for each of the directions, then we will observe that the north–south along with the southwest-northwest direction exhibited an insignificant amount of solar resource variability and the east–west with

Table 6.5 Variability in southeast-northwest direction

FID	Lat	Long	GHI(kWh)	Var. GHI
0	76.85	17.15	5.635815342	0.042826107
12	77.4222	16.5734	5.782713973	−0.104072523
13	76.95	17.05	5.646905479	0.03173597
47	77.05	16.95	5.661399452	0.017241997
48	77.35	16.65	5.780139726	-0.101498277
100	76.3923	17.6009	5.614713425	0.063928025
101	76.65	17.35	5.610085205	0.068556244
102	76.75	17.25	5.627846027	0.050795422
132	77.15	16.85	5.713450959	-0.034809509
137	77.25	16.75	5.747038356	-0.068396907
159	76.45	17.55	5.625354247	0.053287203
173	76.55	17.45	5.63287726	0.045764189
Mean			**5.673194954**	**0.005446495**

Table 6.6 Synoptic view of mean variability for each direction

Sl. No	Directions	Variability (kWh/m2/day)
1	East–West	0.019781
2	North–South	0.00461285
3	Northeast-Southwest	0.017817539
4	Southeast-Northwest	0.005446495

northeast-southwest direction showed a significant amount of solar resource variability. A comprehensive account of variability for each of the directions has been explained in Table 6.6.

2. ***Temporal Variability Results***

(a) *Decadal Versus Annual solar insolation variations.*

Every year, Fig. 6.13 depicts the decadal fluctuation of solar insolation levels. It produced years with solar insolation minimums, maximums, and abnormal fluctuations. The graph clearly shows that the years 2006 and 2010 had relatively high and low variability, respectively, when compared to the rest of the decade. Every year, the linear patterns of variability were also depicted in the diagram. Aside from the picture, Table 6.7 describes the yearly solar insolation values from 2004 to 2014, as well as decadal mean solar insolation values and annual insolation variations for decadal mean solar insolation values.

We can see from Table 6.7 that the years 2007, 2008, and 2010 all had decreasing trends in solar variability. It may be deduced from this year's variability numbers that reported years had steady solar radiations since solar insolation variability was insignificant or extremely low.

(b) *Variations in Annual Versus Monthly Solar Insolation*

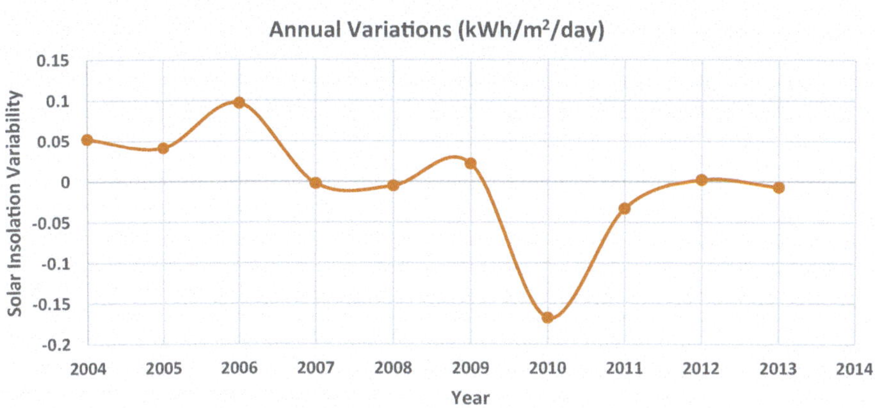

Fig. 6.13 Illustration of annual variability of global solar radiation

Table 6.7 Yearly global solar insolation from 2004 to 2014

Year	GHI (MJ)	GHI (KWh)	Decadal (Mean)	Annual variation
2004	19.17166	5.325460406	5.378145121	0.052684715
2005	19.21178	5.336604833	5.378145121	0.041540288
2006	19.01217	5.281159217	5.378145121	0.096985904
2007	19.36968	5.380467456	5.378145121	0.002322335
2008	19.38042	5.383450792	5.378145121	0.005305671
2009	19.28299	5.356386917	5.378145121	0.021758204
2010	19.96527	5.545908839	5.378145121	0.167763718
2011	19.48132	5.411478175	5.378145121	0.033333054
2012	19.35279	5.375776083	5.378145121	0.002369038
2013	19.38513	5.384758497	5.378145121	0.006613376
*2014**	*20.17434*	*5.603982883*	*5.378145121*	*0.225837762*

* *The values for the year 2014 were available up to July 2014*

Variability Analysis for the year 2004

Every month, variability was measured during the year 2004. The statistics in Table 6.8 show that there are daily variations in the amount of solar insolation, which may be better understood by looking at Fig. 6.14. These variations can be seen throughout each month. During the process of computing the solar insolation variability using the annual mean data, it was observed that there were irregularities in the patterns of the monthly mean values for the year 2004. The selected year's monthly mean values were analysed and compared to the same year's yearly mean value. We can see from the same graph that March has low variability and August has larger variability when compared to the yearly mean values. The results of this investigation were represented in Fig. 8, and the data were tallied.

Variability Analysis for the year 2006

Similarly, a research study on the fluctuation of solar insolation was carried out in the year 2006 to assess the level of variability. The annual mean of the monthly mean values was computed, and the findings were compared to the monthly mean values of the individual months that comprised the same year. The months with the highest, lowest, and most unusual fluctuation were identified. The months with the steady or unstable solar insolation, as well as daily solar insolation changes, were also described in this study. In Table 6.9 and Fig. 6.15, the variability for the year 2006 is depicted. Based on the data shown in the table and the image, the month of April has the least amount of variation, while June, July, and August have the greatest. The amount of solar insolation that varies over November is likewise rather low.

Table 6.8 Values of monthly solar insolation for the year 2004

Months	GHI (MJ)	Yearly Mean (MJ)	Variability (MJ)	Variability (kWh)
Jan	18.35717516	19.18123159	0.82405643	0.228904564
Feb	21.37792908	19.18123159	2.19669749	0.610193747
Mar	24.72104822	19.18123159	5.53981663	1.538837953
Apr	21.86309002	19.18123159	2.68185843	0.744960675
May	20.25424995	19.18123159	1.07301836	0.298060656
Jun	16.92206847	19.18123159	2.25916312	0.627545311
Jul	15.86756793	19.18123159	3.31366366	0.920462128
Aug	14.57105451	19.18123159	4.61017708	1.280604744
Sep	19.11068289	19.18123159	0.0705487	0.019596861
Oct	19.95976866	19.18123159	0.77853707	0.216260297
Nov	17.93982133	19.18123159	1.24141026	0.344836183
Dec	19.23032289	19.18123159	0.0490913	0.013636472

Fig. 6.14 Variability of global solar radiation for the year 2004 on a monthly timescale

Variability Analysis for the year 2010

Following the completion of the 2004 and 2006 variability investigations, for the year 2010, we started research on variability analysis. The results of the investigation into variability indicated that the quantity of solar insolation variation was at its lowest during February, March, and April, while it was at its maximum during July and November. The outcomes of the investigation are summed up in Fig. 6.16 and Table 6.10, respectively. According to the findings of the research, the months of February, March, and April are the ones that have the least amount of variability in solar insolation.

Table 6.9 Monthly solar insolation values for the year 2006

Months	GHI (MJ)	GHI (kWh)	Yearly Mean (kWh)	Variability (kWh)
Jan	19.91936859	5.533157942	5.290072271	0.243085671
Feb	22.77534977	6.326486047	5.290072271	1.036413776
Mar	21.94601883	6.096116342	5.290072271	0.806044071
Apr	24.30869257	6.752414603	5.290072271	1.462342332
May	21.48683633	5.968565647	5.290072271	0.678493376
Jun	17.21105126	4.780847572	5.290072271	0.509224699
Jul	14.21479699	3.948554719	5.290072271	1.341517552
Aug	13.86324073	3.850900203	5.290072271	1.439172068
Sep	18.30674551	5.085207086	5.290072271	0.204865185
Oct	18.94771208	5.263253356	5.290072271	0.026818915
Nov	16.86903411	4.685842808	5.290072271	0.604229463
Dec	18.68227532	5.189520922	5.290072271	0.100551349

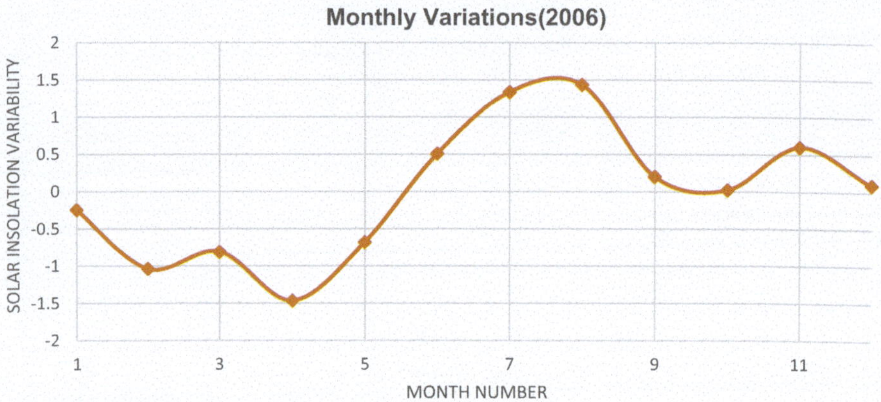

Fig. 6.15 Trends of annual variability of global solar radiation for the year 2006

Variability Analysis for the year 2013

In the end, research on the variability of solar insolation was carried out for the year 2013 to estimate the amount of monthly variation that occurred during that year. The findings are shown in Table 6.11 and Fig. 6.17, respectively. This table and figure demonstrated that solar insolation variability was low throughout February, March, April, and May; nevertheless, the month of March had the lowest fluctuation, which suggests that solar insolation was consistent over that year. In contrast, the most variation occurred during June, July, and August; nonetheless, June and July were the months that had the greatest degree of volatility over the whole year.

The studies of monthly variability that were conducted for the years 2004, 2006, 2010, and 2015, among others, provided us with a synoptic picture of the different

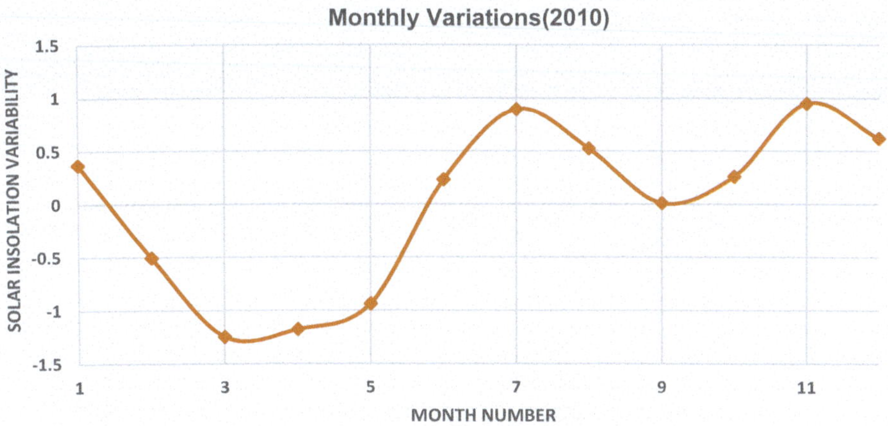

Fig. 6.16 Annual variability of global solar radiation for the year 2010

Table 6.10 Monthly solar insolation values for the year 2010

Months	GHI (MJ)	GHI (kWh)	Yearly mean (kWh)	Variability (kWh)
Jan	18.6484229	5.180117472	5.550007756	0.369890284
Feb	21.79544763	6.054291008	5.550007756	0.504283252
Mar	24.46350388	6.795417744	5.550007756	1.245409988
Apr	24.21881298	6.72744805	5.550007756	1.177440294
May	23.35756408	6.488212244	5.550007756	0.938204488
Jun	19.14795221	5.318875614	5.550007756	0.231132142
Jul	16.76029979	4.655638831	5.550007756	0.894368925
Aug	18.09536823	5.026491175	5.550007756	0.523516581
Sep	19.94039231	5.538997864	5.550007756	0.011009892
Oct	19.03628753	5.287857647	5.550007756	0.262150109
Nov	16.55267221	4.597964503	5.550007756	0.952043253
Dec	17.74361129	4.928780914	5.550007756	0.621226842

patterns of solar insolation fluctuation monthly. The findings of this research, which identified the months with low, high, and anomalous variability, may prove to be of great benefit to academics and planners working on solar resource project planning. This research can give information regarding steady, unstable, and unpredictable solar isolation periods. As a result, whenever someone plans for any form of the solar resource, they will include these "dark periods" or "uncertain or unstable periods," which will improve project planning and execution efficiency. We introduced a new term in the previous clause, which we referred to as "dark periods or uncertain or unstable periods." These times might also be referred to as "periods of solar insolation uncertainty or absence." All the findings of the analysis were plotted on the chart in

Table 6.11 Monthly solar insolation values for the year 2013

Months	GHI(MJ)	GHI(kWh)	Yearly Mean(kWh)	Variability(Kwh)
Jan	19.75579395	5.487720542	5.390500999	0.097219543
Feb	21.39272936	5.942424822	5.390500999	0.551923823
Mar	24.61234751	6.836763197	5.390500999	1.446262198
Apr	25.07176842	6.964380117	5.390500999	1.573879118
May	22.60046912	6.277908089	5.390500999	0.88740709
Jun	14.80750036	4.113194544	5.390500999	1.277306455
Jul	11.62945752	3.230404867	5.390500999	2.160096132
Aug	16.40916037	4.558100103	5.390500999	0.832400896
Sep	20.95838786	5.821774406	5.390500999	0.431273407
Oct	18.22633184	5.062869956	5.390500999	0.327631043
Nov	18.37042747	5.102896519	5.390500999	0.28760448
Dec	19.03526937	5.287574825	5.390500999	0.102926174

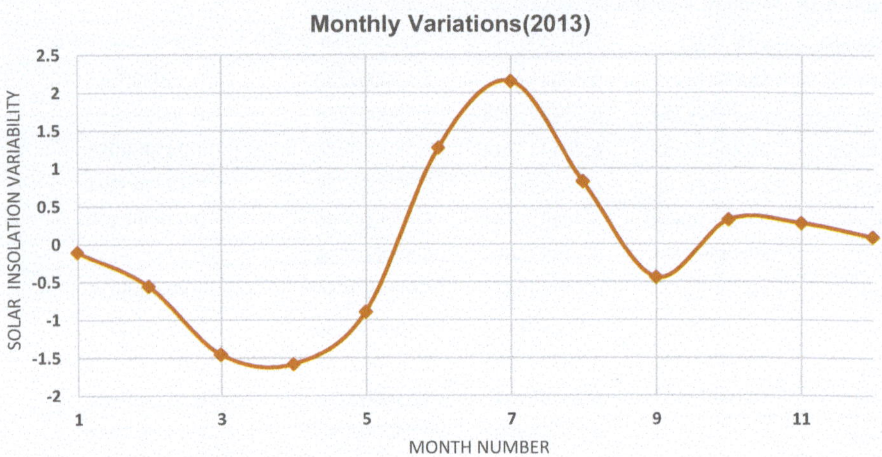

Fig. 6.17 Annual variability of global solar radiation for the year 2013

the relevant places, and the necessary number of explanations was included (wherever required).

6.5.2 *Outcomes from Hyper-Temporal Variability Analysis*

Because of the effects of climate change on a worldwide scale, it is now more vital than ever before to have correct information on the resources that are at our disposal. It is important to have accurate information on the availability of resources and the variability of those resources available on a micro- or hyper-temporal scale. The present defined approach, dubbed "hyper-temporal analysis," was used to investigate the variability of solar insolations for the pilot research in this context.

The hyper-temporal analysis was performed according to the local seasons of the research. Additionally, the analysis was performed for the most crucial months of the season to identify the variability on a microlevel. Daily graphs were produced based on the data on the variability. The results were able to show reasoning and comprehension of the many types of solar insolation. An investigation into a hyper-temporal perspective was carried out in the years 2010 and 2013. The year 2010 had the fewest number of variations and anomalies in the yearly solar insolation variability, whereas the year 2013 was quite like the year of the research on a decadal level. Additionally, the year 2006 had a greater number of yearly variabilities on an annual basis; nevertheless, this year was excluded from your current research since it occurred too far in the past to be relevant to the current year of study. Considering the year 2006 might result in some fascinating discoveries.

Three distinct seasons occur locally in the location where the study was conducted: summer, winter, and monsoon. The objective of the analytical research was to provide an explanation for the characteristics of the changing solar insolation daily for the most significant months of each local season class. To accomplish the goals of this investigation, we focused on the years 2010 to 2013 and examined the months of May, July, and December (one significant month from each season class). Because of the findings of the Hyper-Temporal variability research, we were able to determine the amount of variability that exists on a daily and weekly basis following the local seasons. It also provided academics with a microlevel understanding of the variability phenomenon, which made it possible for them to discover solutions to a range of previously unanswered questions. The results of the "hyper-temporal analysis" were shown as graphs or charts for the audience to peruse (explaining the trends of solar insolation variability daily or weekly basis).

(a) ***Analysis of Daily and Weekly Solar Insolation Variability for the year 2010***

Figure 6.18 presents a visual representation of the various solar insolation that occur on a daily and weekly basis. For this investigation, the month of December in the year 2010 was chosen. December is part of the winter season according to the classification of local seasons. The patterns in solar insolation shown in Fig. 6.18 revealed that the second week of December had the largest variation, whereas the other weeks of the month's solar insolation trends did not display any notable variability.

This also demonstrated that the month is stable about the amount of solar insolation that it receives. The daily fluctuations in solar insolation indicated that the most significant shifts took place on December 6, December 7, and December 9. If these three days from December 2010 are removed, the other days appear to be steady. Figure 18 depicts the visual interpretations.

The second portion of the 2010 analysis was carried out during the monsoon season in July. The variation of solar insolation was analysed on a daily as well as a weekly timescale in this study. The results of the investigation may be seen in Fig. 6.19. The fact that there was more interweekly solar insolation fluctuation may

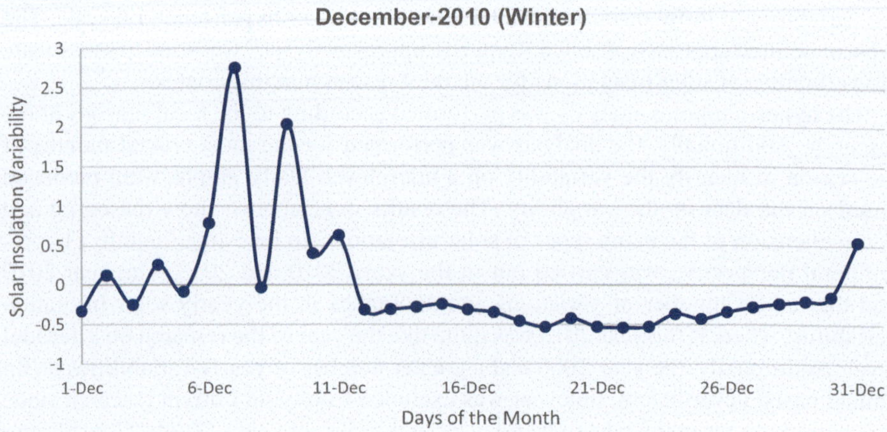

Fig. 6.18 Variability of solar insolation for December 2010 (Winter Season)

be able to be noticed in the earliest emergence of solar insolation variability patterns. The solar variability trends in the figure demonstrated that the maximum level of fluctuation occurred throughout all weeks of July in the year 2010 and that there was no significant consistency in the variability of solar insolation trends outside of these weeks. In terms of solar insolation fluctuation, this also spoke to the fact that months had very unpredictable weeks. In a similar vein, the research that looked at the variability of solar insolation patterns daily found that July had practically every day with the largest variability. The findings of the study that was conducted daily throughout July 2010 revealed that the days with the greatest variation in solar insolation were July 3, July 6, July 7, July 24, July 27, July 28, and July 30, while the other days in the month exhibited only minor variation. It is possible to utilize Fig. 6.19 to have a better grasp of the data by visually representing all the different solar variability patterns.

Finally, in the year 2010, the analysis for May was completed. Figure 6.20 depicts the outcomes of the analysis. The analysis was carried out on a weekly and daily basis. Weekly diurnal fluctuations in solar insolation were recorded in the weekly analysis. In May 2010, there were complicated variances in solar insolation trends. The research found that all weeks in May had the most fluctuation and that there was no significant stability in the solar insolation patterns outside of these weeks. The fact that the month is particularly unstable in terms of solar insolation variations was discussed in this way. The findings of the daily study for May 2010 revealed that the biggest variability of solar insolation occurred on May 2, May 8, May 18, May 19, May 20, and May 22, while the remainder of the month had minimal variability. Even if we exclude the above-mentioned days from May 2010, the other days were not determined to be particularly steady. The dynamics of weekly and daily solar insolation fluctuation patterns are shown straightforwardly in Fig. 6.20.

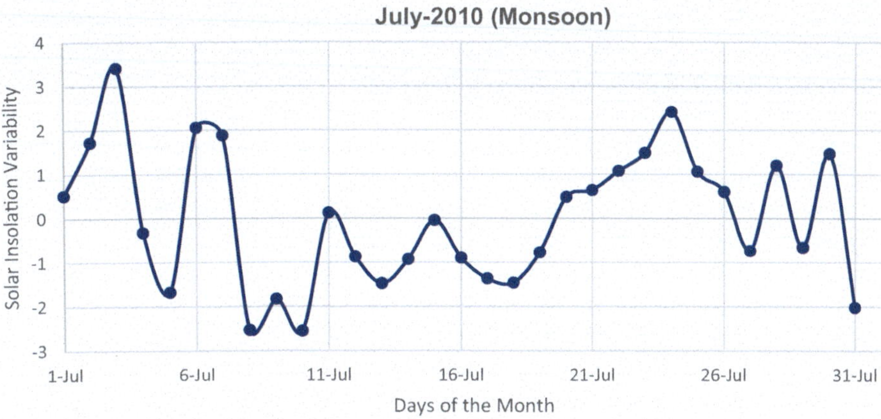

Fig. 6.19 Variability of solar insolation variability for July 2010 (Monsoon season)

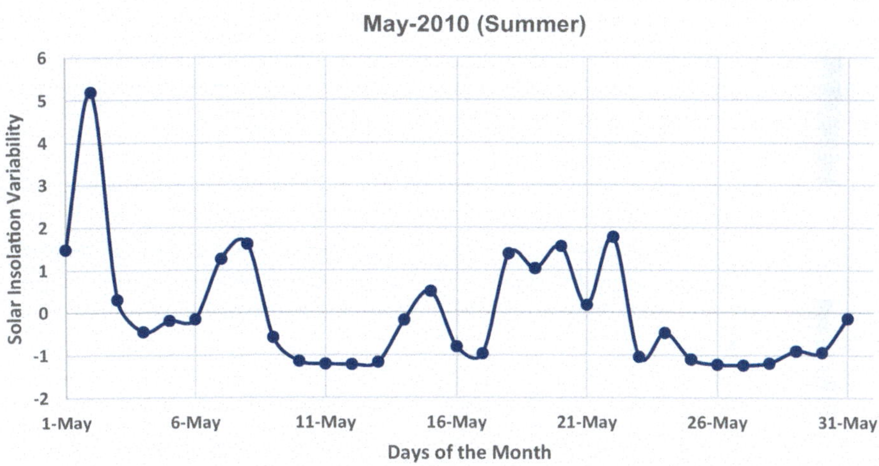

Fig. 6.20 Variability of solar insolation for May 2010 (Summer season)

(b) ***Daily and Weekly Variability Analysis of Solar Insolation for the year 2013***

Figure 6.21 depicts the results that were uncovered by the variability study carried out for the year 2013 specifically. The study was carried out on both a weekly and a daily basis during December 2013, inclusive. Based on the findings of the weekly research, it was determined that, except for the first week of December 2013, each of the subsequent weeks had very modest variations. It can be seen in Fig. 6.21 that the second, third, and fourth weeks of the month had patterns of variation that were quite like one another. On the other hand, there was no notable variability in the solar insolation trends for any of the other weeks of the month. This proved that the month is essentially consistent in terms of the solar insolation that it receives. According to the patterns of solar insolation fluctuation daily, all the remaining days of the month, except December 2nd, exhibited modest changes in the solar insolation levels. Therefore, if the day in question from December 2013 is removed from consideration, it is presumable that the other days will not change. Figure 6.21 depicts the visual interpretations for your reference.

The second portion of the analysis for 2013 consisted of looking at the datasets from July of that year. Daily and weekly studies of the solar insolation's varying conditions were carried out for this variation research. The results of the examination are shown in Fig. 6.22. The first indications of trends showed that the interweekly variation in solar insolation was present to a higher degree over the whole of the month. The trends in solar variability also indicated that the greatest level of fluctuation took place throughout the entire month of July 2013 and that there was no significant consistency in the variability of solar insolation trends outside of these weeks. This was indicated by the fact that the highest level of fluctuation took place. These also explored the fact that months had extremely erratic weeks in respect of the amount of solar insolation that they received. In a similar vein, the daily variability research of solar insolation trends discovered that practically every day in July had the largest fluctuation. The outcomes of the research that were conducted daily during July 2013 indicated that July 16, July 17, July 18, July 19, and July 31 had the greatest amount of variation in solar insolation, while the other days of the month had the least amount of variation. For a better understanding of the data, Fig. 6.22 may be used to visualize all the solar variability patterns.

Finally, in the year 2013, the analysis for May was completed. Figure 6.23 depicts the outcomes of the analysis. On a weekly and daily basis, the analysis was carried out. Weekly diurnal fluctuations in solar insolation were recorded in the weekly analysis. For May 2013, there were composite aberrations in solar insolation patterns. According to the findings of the study, the solar insolation variation throughout the other weeks of the month was insignificant, except for the third and fourth weeks of the month. Additionally, the researchers discovered that the solar insolation variability tendencies were not particularly steady. Because of this, the month may be described as stable in terms of the fluctuation of the solar insolation, with the greatest number of weeks displaying a negative variation. Only May 15, May 17, and May 31

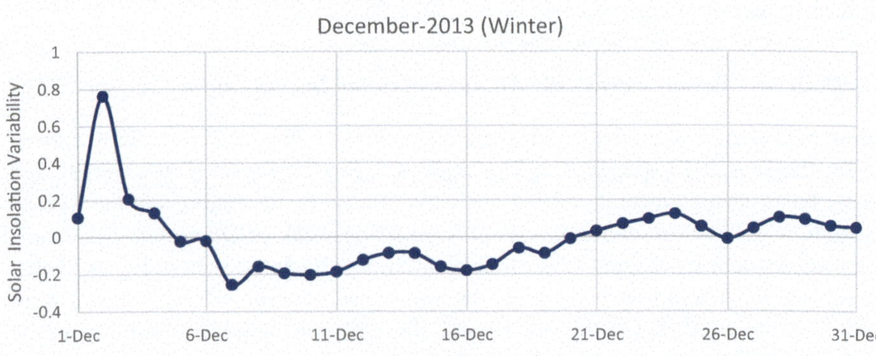

Fig. 6.21 Solar insolation variability for December 2013 (Winter season)

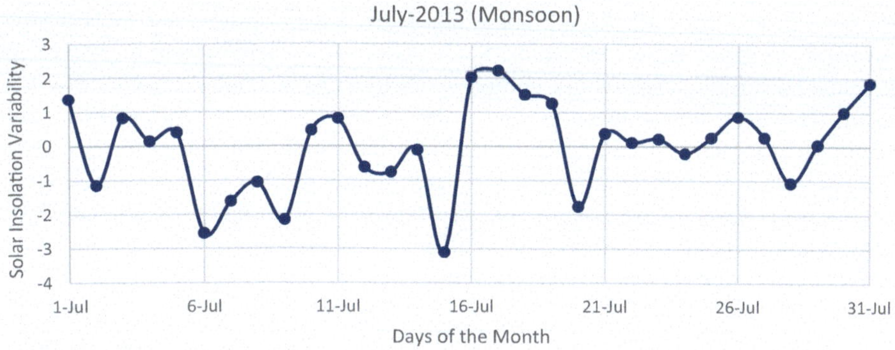

Fig. 6.22 Variability of solar insolation for July 2013 (Monsoon season)

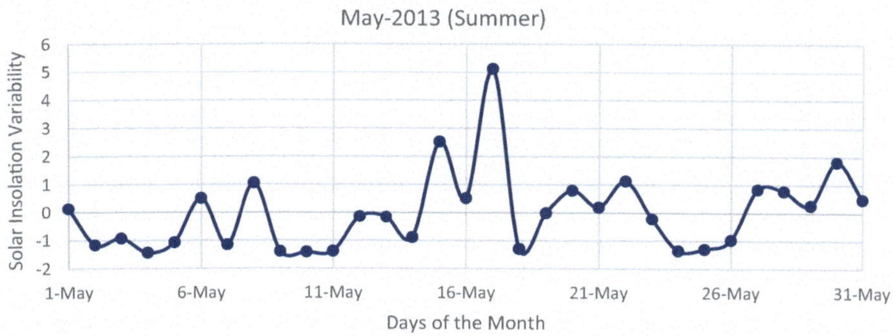

Fig. 6.23 Variability of solar insolation for May 2013 (Summer Season)

showed the greatest variance in sun insolation during the whole month of May; the other days of the month showed only modest fluctuations in solar insolation levels. The results of the monthly research revealed this information. When these days are removed from the month, the remaining days are determined to be marginally stable days. Fig. 6.23 depicts the dynamics of weekly and daily solar insolation fluctuation patterns.

The hyper-temporal analysis of data given here gave us concepts for reporting on solar insolation variability, which might help our solar resource planner analyse variation trends on a monthly, weekly, and daily basis.

6.6 Summary

Due to a lack of traditional energy sources, it is now difficult to ensure an uninterrupted electrical supply. To meet our constantly expanding energy demands, it has become necessary to investigate alternative energy sources. As a result, the researcher has been tasked with identifying some alternative energy-producing sources that can meet our future energy demands. Solar photovoltaics have shown to be one of the most powerful alternatives or complements to traditional energy sources in this regard. However, before installing any solar energy harvesting equipment in any region. It is necessary to assess the viability or potential of that location for solar energy production. During the process of determining whether or not an area is suitable for the production of solar energy, one of the most important factors to consider is the "spatio-temporal and hyper-temporal variability study of solar insolation." This problem is now the subject of much dispute.

Solar energy, unlike conventional electrical power generation (e.g., fossil, or nuclear), is unreliable. The weather, as well as the cycle of days and seasons, influence the production of a solar power plant. The data were analysed on the necessary geographical and temporal scales in this study. The variability statistics reported in this study will help you understand the scope of observations needed to accurately define a solar resource for a specific application. A shorter local measurement time may suffice in places with minimal interannual variability and vice versa. Which of these will assist the analyst in gaining greater trust in a data set for a certain analysis?

These findings revealed that the solar resource has a wide range of fluctuation, with values ranging from inconsequential to extremely substantial. Solar insolation variability in the Kalaburagi area was found to be mostly driven by topography slopes and orientations, which cause significant fluctuations in the city's eastern half. For the Kalaburagi district, our study validated a spatio-temporally explicit technique for tracking the heterogeneity of solar insolation fluctuations.

References

Ahmad F, Goparaju L (2018) A geospatial analysis of climate variability and its impact on forest fire: a case study in Orissa state of India. Spat Inf Res 26(6):587–598. https://doi.org/10.1007/s41324-018-0197-8

Ajayi O, Ohijeagbon O, Ajanaku K, Aasa S, Omotosho O (2016) Techno-Economic Assessment of Renewable Hybrid Systems for Rural Electrification and Distributed Generation in Selected Sites across Nigeria. Journal of African Research in Business & Technology 2015:1–31. https://doi.org/10.5171/2015.124767

Ali, T., Nahian, A. J., & Ma, H. (2020). A hybrid multi-criteria decision-making approach to solve renewable energy technology selection problem for Rohingya refugees in Bangladesh. *Journal of Cleaner Production, 273*. https://doi.org/10.1016/j.jclepro.2020.122967

Antoszewski, P., Krzyżaniak, M., & Świerk, D. (2022). The Future of Climate-Resilient and Climate-Neutral City in the Temperate Climate Zone. *International Journal of Environmental Research and Public Health, 19*(7). https://doi.org/10.3390/ijerph19074365

Banzhaf, E., Wu, W., Luo, X., & Knopp, J. (2021). Integrated mapping of spatial urban dynamics-a european-chinese exploration. part 1-methodology for automatic land cover classification tailored towards spatial allocation of ecosystem services features. *Remote Sensing, 13*(9). https://doi.org/10.3390/rs13091744

Bhakta S, Mukherjee V (2017) Performance indices evaluation and techno economic analysis of photovoltaic power plant for the application of isolated India's island. Sustainable Energy Technol Assess 20:9–24. https://doi.org/10.1016/j.seta.2017.02.002

Borfecchia F, Caiaffa E, Pollino M, Cecco LD (2014) Remote Sensing and GIS in planning photovoltaic potential of urban areas. European Journal of Remote Sensing 47(195):195–216. https://doi.org/10.5721/EuJRS20144713

Busho SW, Wendimagegn GT, Muleta AT (2021) Quantifying spatial patterns of urbanization: growth types, rates, and changes in Addis Ababa City from 1990 to 2020. Spat Inf Res 29(5):699–713. https://doi.org/10.1007/s41324-021-00388-4

Ellis AW, Hilderbrandt ML, Thomas WM, Fernando HJS (2000) Analysis of the climatic mechanism contributing to the summertime transport of lower atmospheric ozone across metropolitan Phoenix, Arizona, USA. Climate Res 15(1):13–31. https://doi.org/10.3354/cr015013

Gambhir, A., Dixit, S., Toro, V., Singh, V., Ramachandra, T. V., Jain, R., Krishnadas, G., Nayak, J. K., Prajapati, J. a., Noel, Cressie;Christopher, K. W., Lean, J. L., Sayigh, A. A. M., Pareek, A., Studios, B., Commission, C. E., National, T., Plan, A., Change, C., Nehru, J., … Observatory, N. (2010). A Guide to PV System Design and Installation. In *Solar Energy* (Vol. 44, Issue 5). Academic Press. https://doi.org/10.1002/wcc.018

Guan, K. (2011). *Surface and ambient air temperatures associated with different ground material: a case study at the University of California, Berkeley*. 14. http://nature.berkeley.edu/classes/es196/projects/2011final/GuanK_2011.pdf

Islam, M. S. (2018). A techno-economic feasibility analysis of hybrid renewable energy supply options for a grid-connected large office building in southeastern part of France. *Sustainable Cities and Society, 38*(January 2018), 492–508. https://doi.org/10.1016/j.scs.2018.01.022

Kestens Y, Brand A, Fournier M, Goudreau S, Kosatsky T, Maloley M, Smargiassi A (2011) Modelling the variation of land surface temperature as determinant of risk of heat-related health events. Int J Health Geogr 10(1):7. https://doi.org/10.1186/1476-072X-10-7

Kumar, D. (2019a). Hyper-temporal variability analysis of solar insolation with respect to local seasons. *Remote Sensing Applications: Society and Environment, 15*. https://doi.org/10.1016/j.rsase.2019a

Ramachandra TV, Krishnadas G, Jain R (2012) Solar Potential in the Himalayan Landscape. ISRN Renewable Energy 2012:1–13. https://doi.org/10.5402/2012/203149

Rn HK, Shafiullah GM, Maung A, Oo T, Stojcevski A (2014) Feasibility study of Standalone Hybrid Power system modeled with Photovoltaic modules and Ethanol generator for Victoria. International Journal of Engineering Science Invention 3(8):54–69

Saeed, U., & Ahmad, S. R. (2021). Emerging GIS based rehearses for assessment of urban environmental sustainability and apposite ranking. *Fresenius Environmental Bulletin, 30*(3), 3047–3058. https://www.scopus.com/inward/record.uri?eid=2-s2.0-85105567443&partnerID=40&md5=15cfcdc5dfe7f67645f469353fe689c9

Snir K, Pearlmutter D, Erell E (2016) The moderating effect of water-efficient ground cover vegetation on pedestrian thermal stress. Landsc Urban Plan 152:1–12. https://doi.org/10.1016/j.landurbplan.2016.04.008

Tanskanen A, Jantunen A, Saksa JM, Partanen J, Bergman J, Bertsch V, Treitz M, Geldermann J, Rentz O, Di Silvio B, Cesarotti V, Introna V, Babatunde OM, Babatunde DE, Denwigwe IH, Adedoja TB, Adedoja OS, Okharedia TE, Kwakwa PA, … Authors F (2021) New directions in UK energy policy. Int J Energy Sector Manage 15(2):157–172. https://doi.org/10.1108/IJESM-12-2019-0001

Wu DD, Olson DL, Birge JR (2013) Risk management in cleaner production. J Clean Prod. https://doi.org/10.1016/j.jclepro.2013.02.014

Xie Z, Wu R, Wang S (2021) How technological progress affects the carbon emission efficiency? Evidence from national panel quantile regression. J Cleaner Prod 307. https://doi.org/10.1016/j.jclepro.2021.127133

Yang J, Huang X, Tang Q (2020) Satellite-derived river width and its spatiotemporal patterns in China during 1990–2015. Remote Sens Environ 247. https://doi.org/10.1016/j.rse.2020.111918

Zhao TF, Fong KF (2017) Characterization of different heat mitigation strategies in landscape to fight against heat island and improve thermal comfort in hot–humid climate (Part I): measurement and modelling. Sustain Cities Soc 32:523–531. https://doi.org/10.1016/j.scs.2017.03.025

Chapter 7
Feasibility Analysis of an Energy System

Abstract A feasibility study is a sensible way for any investor to find out if a project they're thinking about funding is likely to fail and produce little to no money. Would they still contemplate investing a single dime? Millionaire investors are divided on this type of informed decision-making. Therefore, smart investors in solar power projects want a solar feasibility analysis before putting a single shovel in the ground. The first thing you should do is do a solar feasibility analysis. A solar feasibility assessment is not necessary before starting a photovoltaic project, of course. It is a person's right to spend several thousand dollars on solar power equipment. However, if their land is not in a good location for solar power, they may not get a return on their investment. They might not even be able to recuperate their equipment expenditures. Furthermore, putting such equipment on a property may be against zoning and permission rules. Investing in solar electricity is not a cut-and-dry proposition, contrary to common opinion. A comprehensive study is necessary, which includes the following due diligence services (a) Financial Modelling—Estimate and validate project ROI, (b) Project Costing—Complete a thorough budget analysis encompassing project, labour, and materials, (c) Site Analysis—Determine whether the area is suitable for solar power projects, (d) Costs of Interconnection—Identify possible interconnection costs and issues, and (e) Engineering Review—Before obtaining permissions, the engineering must be thoroughly reviewed. Whatever the results of the solar feasibility study reveal, one thing is certain: investing without first learning more is not just a risk, but it may also be disastrous. A recent study on renewable energy resources is one of the keys aims for recent improvements in big centres to make them less reliant on imported energy by developing and producing solar power locally. This section introduces the smart grid methodology, which is used to examine the viability of smart solar grid system design.

Keywords *Feasibility analysis* · *Smart solar grids* · *Power generation*

7.1 Overview

Since the 1950s, when photovoltaic cells were first developed, they have been included in the design of spacecraft. Despite this, a progressively increasing terrestrial

D. Kumar, *Renewable Energy Scenarios in Future Indian Smart Cities*, Advances in Geographical and Environmental Sciences, https://doi.org/10.1007/978-981-19-8456-3_7

industry came into existence during the energy crisis of the early 1970s. Even though solar photovoltaics were only touched on briefly in the first chapter, a review is in need. In solar photovoltaic, also known as SPV, technology, semi-conducting devices known as solar cells are used to directly convert sunlight into energy (Farhangian Marandi et al. 2018; Lewis et al. 2005). A photovoltaic module is constructed from solar cells that are connected and hermetically sealed. When building bigger systems and power plants, solar modules are often combined with additional components like storage batteries. Photovoltaic (PV) systems and power plants are designed from the ground up to be very dependable and versatile in how they run (Islam 2018; Li et al. 2020; Rybach. 2014).

Following huge rises and volatility in fossil fuel costs, as well as the environmental requirement of lowering greenhouse gas emissions to decrease the rate of global climate change, interest in renewable energy technology has grown (Aleksandrowicz et al. 2019; Cooper et al. 2022; Tsoutsos et al. 2005). Increased and sustained deployment of novel technologies such as solar energy is heavily reliant on the economics and reliability of the solar systems installed throughout the market entrance phase. The cost-effectiveness and dependability of a system are the results of a well-planned project based on correct information on the solar resource available (Borfecchia et al. 2014; Gils et al. 2017; Zhang et al. 2019). Smart grids are designed to save energy by adjusting supply to demand. The implications of solar penetration on the electrical grid, as well as solutions for reducing variability, are examined. The term "smart grid" is not universally defined. It is dependent on the smart grid system's development goal. A smart solar grid (SSG) is a highly efficient energy system that includes dependability, flexibility, centralized and distributed power generation, and a completely automated power network (Alizadeh et al., 2016; Dovì et al., 2009; Nayan et al., 2013). This research examines smart grid business cases, present interests, and infrastructures to make preliminary suggestions for a smart grid project and the following steps in implementation. The feasibility analysis is the final section of the research. A feasibility study examines the practicality of a proposal, as the name indicates. The feasibility study is designed to answer the question, "Should we go through with the suggested project idea?" The goal was to assess the solar photovoltaic system's potential as well as the economic and environmental implications. The research tried to analyse the importance of smart grid installation in a pilot region and conduct a full feasibility analysis.

7.2 Sub-objectives

1. Exploring smart solar grids, their potential, and existing technologies.
2. Feasibility analysis of smart solar grids for proficient power generation.

7.3 Workflow Overview

Figure 7.1 depicts the general approach process used in this portion of the investigation. To make the task more manageable, it was divided into three sections. In Fig. 6.1's process, we can see that feasibility is the study's major issue, which is then separated into three key components as shown in Fig. 7.2, which describe the many analyses used to acquire the findings.

These four primary processes may be stated as follows: (1) investigate the possibilities for smart solar grid deployment, (2) collect and analyse data, (3) determine the viability of smart solar grids for efficient power generation, and (4) give inputs for future power-generating projects.

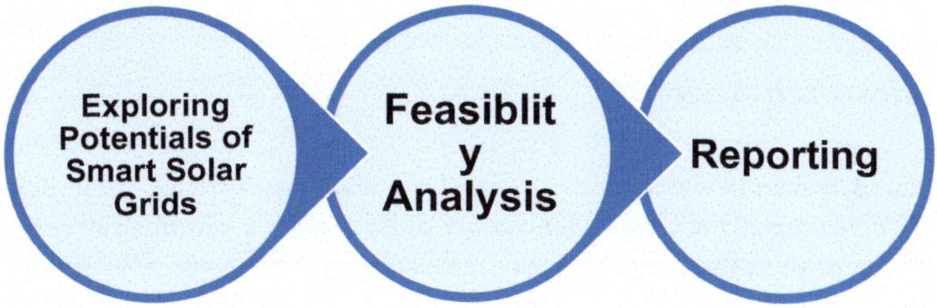

Fig. 7.1 Schematic for workflow diagram

Fig. 7.2 Layout for feasibility analysis

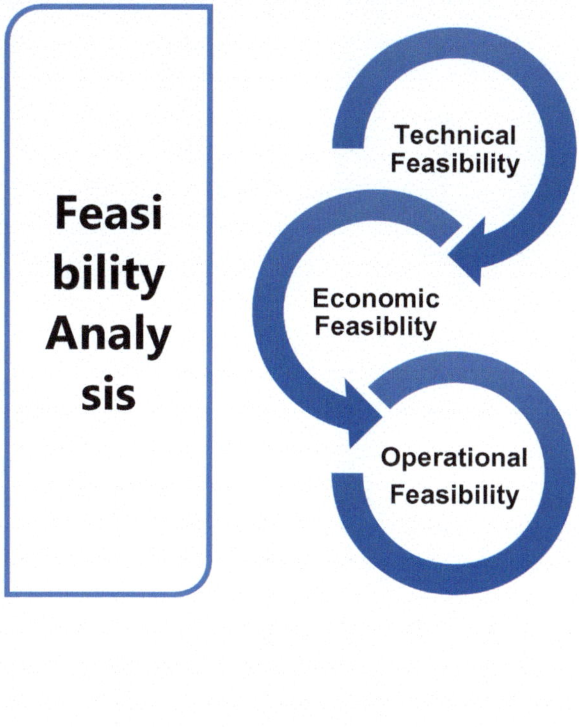

7.4 Methods

This section covers the approach, data sources, and software/tools utilized to accomplish the goal. The analytical research approach is used in the presentation component of the work. The estimation and selection of procedures used in data analysis are referred to as the analytical research method. The analytical design is an important part of study design since it allows us to choose how the data will be evaluated as well as specifics on the consequences of the data/observation and the final output. Mathematical and statistical models/algorithms are also used in analytical procedures. Analysing the situation and understanding the connection between information that has been visually interpreted or digitally processed and the actual situation or events that are taking place on the ground is also an analytical method (Fig. 7.3).

7.4.1 *Research Area*

Experimental research was carried out in the Kalaburagi district, as indicated in Fig. 6.3. We've chosen the Central University of Karnataka's administrative building in Kalaburagi as our pilot research area for this phase of the project. With 654 acres

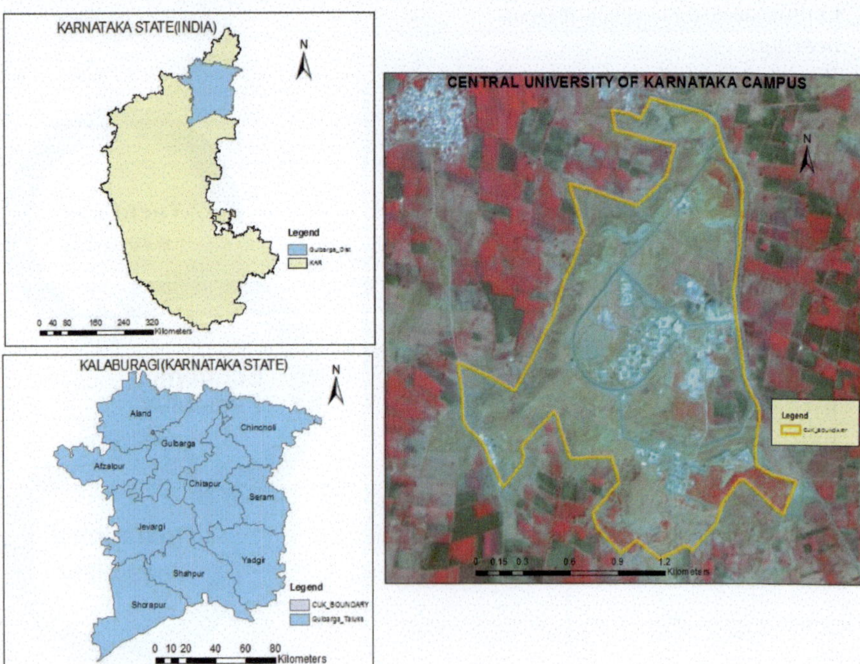

Fig. 7.3 Location of the study area

of land, the Central University of Karnataka (CUK) Campus is located on the north-western side of Kalaburagi city (on the outskirts of Kalaburagi city, about 21 miles). The geographical area is largely constrained to the latitudes of 17°41' N and 17°44' N, and the longitudes of 76°65' E and 76°68' E. This location was used to conduct a feasibility assessment on solar grids that had already been installed.

7.4.2 Datasets Used

GCPs from handheld GPS, Survey of India topographic sheet on a scale of 1:50 K, SRTM DEM/BHUVAN DEM, and very high-resolution (VHR) satellite data from Google Earth were used in this study (site validations). Other free archive data sets for execution and validation from BHUVAN NOEDA (NRSC), Landsat Archive & Earth Explorer (USGS).

The data were acquired from a variety of open source or educational sources, as well as on-the-ground data. Additional information was gathered from books, journals, publications, and websites. The following are some of the many information sources mentioned:

1. Online gateways and databases *(ResearchGate).*
2. Ministry of New and Renewable Energy *(Website, Reports, Manuals)*.
3. Online Journals *(e.g., Solar Energy, Energy, Journal of Solar Energy Engineering).*
4. Books in the context of infrastructure management, trends in energy sciences.
5. Reference texts from ESRI, ERDAS Inc. publications and manuals.

7.4.3 Software and Tools Used

1. Energy Plus 8.0
2. Geospatial Toolkit *(NREL)*
3. System Advisor Model *(SAM)*
4. HOMER *(Hybrid Optimization of Multiple Energy Resources)* Software
5. RET Screen Plus Software
6. ESRI ArcGIS 10. x (*for vector data analysis*).

7.4.4 Smart Solar Grids Potentials

Our smartphones, along with the vehicles we drive, the appliances we use, and the homes we live in, are all getting more intelligent. The moment has come for the electric grid to get up to speed. The term "electric grid" refers to the enormous network of transmission lines, substations, and power plants that supplies our homes as well as our places of business with the electricity they need to function. In the process of transmission and distribution, we lose around 7% of the power that we create. Many power plants, which are often the dirtiest, least efficient, and most expensive to operate, are only utilized to produce electricity for a few dozen hours each year since their primary purpose is to meet the peak demand for energy. These so-called Peaker plants run houses at a financial loss while also contributing to environmental pollution. However, in comparison to our current electric grid system, which is inefficient, expensive, and a significant contributor to pollution, the smart grid represents a more modern, more efficient, and cleaner alternative.

The electric grid has been a one-way channel from power plants to consumers ever since Thomas Edison turned on the world's first commercial generator in New York City in 1882. This event marked the beginning of the modern age of commercial electricity generation. Since the beginning of time, we've been using this system, this has always been the case. Because of the implementation of a smart grid, which incorporates computer intelligence and communications technology, the energy distribution network, and a significant portion of what is attached to it, such as solar panels, smart appliances, and plug-in automobiles, may now function as a two-way street. This is one of the potential benefits of the smart grid. It offers a lot of promise in terms of producing power that is safer, more efficient, has better air quality, and reduces emissions of greenhouse gases. Figure 7.4 provides an in-depth explanation of the steps that must be taken to develop the most advanced smart solar grid technology.

On a smart grid, residences and places of business are in constant contact with the grid to determine the mode of operation that is both the most cost-effective and the most efficient. Consumers may track their energy consumption using smartphones or PCs, and digital smart meters give real-time electrical data, replacing conventional meters with spinning discs. To achieve carbon neutrality, solar-powered homes should produce the same amount of electricity as is used within the home. This should result in no more carbon dioxide being released into the atmosphere. There are a few

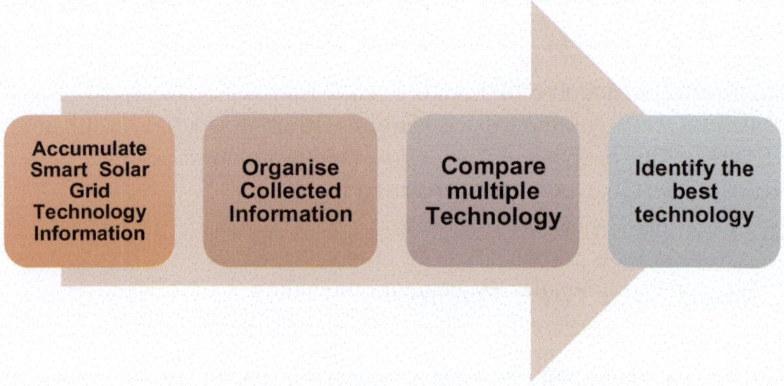

Fig. 7.4 Flowchart for identifying the best smart solar grid technology

possible courses of action, one of which is to construct a "green" and intelligent grid that minimizes waste while maximizing productivity.

As the global movement towards energy efficiency and sustainability gathers steam, the Research Triangle Area (RTA) is seeing an uptick in both interest in and demand for programs about smart grids. The present market for applications with SSG cells may be broken down into three different categories: applications for professionals, those living off the grid, and those linked to the grid.

Professional sector

This industry includes some of the oldest types of exploitation, such as distant telecommunications repeater stations, as well as some of the most cutting-edge technology, such as the US government's intention to build a 10-gigawatt solar power plant that orbits the earth.

Off-grid or stand-alone System

Stand-alone power generation systems are those that are not connected to the public utility grid. In certain off-grid areas, such as those that are less than a quarter mile distant from the power lines, it may be more cost-effective to build a stand-alone photovoltaic system than to extend the power lines. They are a wonderful option for use in distant regions that are sensitive to the surrounding environment, such as national parks, cottages, and rural residences since they do not harm the natural habitat. In rural locations, small solar arrays that are completely self-contained are often constructed to generate power for farm lights, electric fence charges, and solar water pumps. This electricity is then utilized to supply animals with drinking water. However, the majority of systems need battery storage to keep the energy that is generated during the day available for use during the night. Direct-coupled systems are the only kind of system that works solely during daylight hours; thus, they do not require any form of electrical storage. Some of these systems, which are collectively referred to as hybrid systems, combine solar energy with electricity generated from other sources of energy, such as diesel or wind.

Grid-connected System

Grid-connected photovoltaic systems sometimes referred to as grid interface systems can sell excess power to the utility company while simultaneously drawing energy from the utility grid when the power supply from the home system is inadequate. These technologies do away with the need for battery storage, which means that even if the planning for grid connection could be difficult, it is no longer necessary. Some utilities may allow a practice known as "net metering," which enables property owners to make money off their extra electricity by selling it to the utility. The price of photovoltaic electricity has decreased by a factor of 10 over the last two decades, and this trend is expected to continue, making the market for this kind of energy even more competitive.

7.4.5 Feasibility Analysis

The activity of the project includes the creation of renewable energy via the use of photovoltaics. This feeds electricity into and/or displaces power from an electrical distribution system. Originally, this power would have been generated by at least one fossil fuel-fired generating unit. The project activity makes use of solar energy, which is one of the most ecologically friendly solutions for addressing the state's energy demands. Furthermore, the project activity enables technology transfer to increase the quality of electricity delivered at the grid's tail end, and it is one of the first of its kind in the region.

The whole sequence of stages for conducting the feasibility analysis of the chosen technology is shown in Fig. 7.5. The project activity is viewed as a pilot or demonstration project to demonstrate the efficacy of using solar energy for grid-connected electricity generation to enhance the quality of power delivered at the grid's tail end. The successful completion of this demonstration project activity will lead to the development of more similar projects in the region, lowering the use of fossil fuels for the generation of equivalent power outputs and achieving the project's goals. The following are the main characteristics of the project activity.

The profile of the solar photovoltaic power plant deployed in the research region is described in Table 7.1. The processes shown in Fig. 6.5 were followed to conduct the project feasibility study. These assessments gave us the information we needed regarding each of the feasibility parameters (which were taken into consideration). These parameters were then used to build the appropriate matrices.

During the period of analysis, a feasibility study needs to be carried out before the highest level of administration decides what course of action to pursue. A corporation may determine whether an information system is feasible by evaluating how useful or applicable it will be to the business. Throughout the system analysis stage of any system development life cycle, three distinct feasibility milestones must be met. A choice will be made on whether the development of the system will proceed when

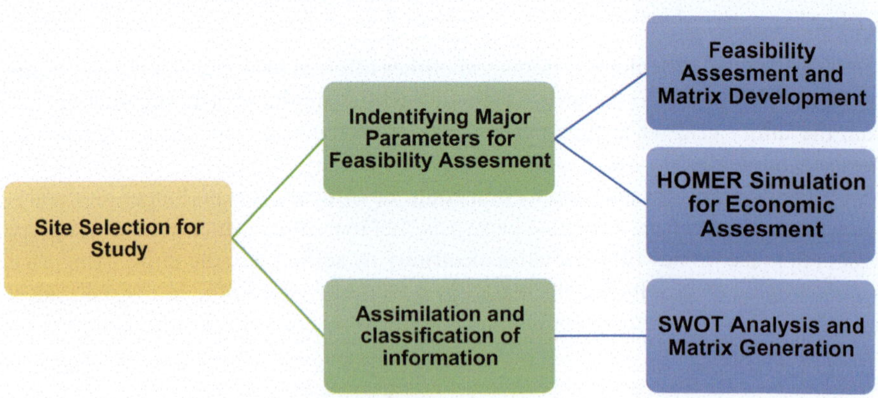

Fig. 7.5 Network diagram for a feasibility study

Table 7.1 Specifications of 40 kW solar energy-electricity system (roof top)

Solar PV plant CUK administrative block		
S. No.	Parameter	Specifications
1	PV array capacity	40k Wp
2	PV module	280 Wp
3	Number of PV modules	152
4	Modules in series	18
5	Number of series in strings	08
6	Number of structure	38
7	Solar inverter	20 KWp (2 Nos)
8	DC distribution box	1 IN/1 OUT (8 Nos)
9	AC distribution box	1
Solar PV specifications		
1	Technology type	Multicrystalline
2	Manufacturer	Central Electronics Limited
3	Model	PM 280
4	Power max (Pmpp)	280 W
5	Maximum system voltage	1000 V
6	Voltage at pump	35.00 V
7	Open circuit voltage	44.00 V
8	Current at Pmpp (Impp)	8.00 A
9	Short circuit current	8.80 A
10	Fuse rating	15A
11	Irradiance (at 25 °C)	1000 W/m^2
12	Field wiring copper	14AWG
Solar inverter specifications		
1	Manufacturer	Sunny tripower
2	Model	STP 20000TLEE-10
3	V_{DC} max	1000 V
4	$V_{DC\ MPP}$	580–800 V
5	I_{DC} max	36 A
6	$V_{AC,r}$	380/400/415 V
7	$P_{AC,r}$	20,000 W

(continued)

Table 7.1 (continued)

Solar PV plant CUK administrative block		
S. No.	Parameter	Specifications
8	S_{max}	20,000 VA
9	$f_{AC,r}$	50/60 Hz
10	I_{AC} max	29A
11	Cos(ϕ)	0.8–1.0

each of these checkpoints is finished. At each of the checkpoints, a feasibility study was performed by conducting the following three tests: an operational feasibility test, a technical feasibility test, and an economic feasibility test.

Scope definition checkpoint

The first checkpoint for the viability of the plan, in which the primary emphasis is on answering questions such as "are the problems and possibilities warrant the expense of a complete assessment of the existing system?" At this step, the validity of the problems and opportunities is evaluated. The expenses of development, timescales, and other considerations are not accounted for in this feasibility study. After analysing the actual challenges and possibilities, the system analyst will provide an estimate of the cost of development.

Problem analysis checkpoint

The detailed measurement and issue analysis of the present system are followed by the second checkpoint for the viability of the project. Once both the possibilities and the problems have been brought to light, it is not difficult for the system analyst to provide a reliable cost estimate for the development of the system. At this point, the bare minimum development cost is the cost of resolving the issues. The true development cost, on the other hand, was determined by the requirement analysis.

Decision analysis checkpoint

Following the completion of the user requirements, a substantial feasibility study will be done at the third feasibility checkpoint. At this moment, several alternative solutions may be chosen. Each of these solutions describes the input/output methods, data storage needs, software and hardware prerequisites, and so on. At this point in the process, each accurate cost estimate for the development may be accessed.

Word	Definition
Feasibility	How useful or effective the system is judged
Economic feasibility	A way to evaluate a project's cost-effectiveness
Net present value	A process for evaluating the yearly expenses and advantages of several options
Operational feasibility	An evaluation of how well a solution satisfies the needs of the organization's specified system for problem-solving

(continued)

(continued)

Word	Definition
Payback analysis	A method is a widely used method for figuring out if and when an investment will pay for itself
Return on investment	A method of the study evaluates the lifetime profitability of different options
Technical feasibility	A way to evaluate the viability of a technological solution as well as the accessibility of technical resources and knowledge

The current trends of the approaches created with the potential to finish a project effectively, taking into consideration legal, economic, technological, scheduling, and other variables, are the focus of the smart grid feasibility study. A feasibility study helps project managers to evaluate the potential bad and positive effects of a project before committing too much time and money. Instead of merely diving in, the research examines a certain notion to make informed conclusions. The research painted a picture of the situations based on the available materials. A feasibility study considers the following factors.

(a) ***Technical Feasibility***

It is a measure of a technological solution's viability as well as the availability of technical resources and experience (keeping in mind the existing technology and available personnel). Technological feasibility is a metric that assesses a technical solution's viability as well as the availability of technical assets and skills. Scientific/Technical feasibility often covers three primary issues:

- Is the technology or solution presented feasible?
- Do we have the requisite technologies right now?
- Do we have the requisite technical knowledge?

(b) ***Economic Feasibility***

It is a metric for determining a project's or solution's cost-effectiveness. It's a method of calculating the advantages and expenses of developing a system. We'll need to compare them to figure this out. There are both real and intangible expenses associated with a system. Items that can be easily measured in terms of money and with certainty are referred to as tangible costs. Hardware expenses, labour costs, and other physical expenditures are all part of the system development process. Intangible costs are expenses incurred as a result of the system that cannot be quantified in terms of money or with certainty. Customer goodwill and employee morale are two examples of intangible expenses.

A system, like expenses, may give a variety of advantages to an organization. There are two types of benefits: tangible and immaterial. A tangible advantage can be quantified in terms of money and is guaranteed. Reducing costs and raising earnings are two examples of tangible advantages. Benefits obtained from the system that cannot be measured in terms of money or with certainty are referred to as intangible benefits. Competitive needs, increased company reputation, and speedier decision-making are examples of intangible advantages. Benefits, in most cases, may enhance profits while lowering expenses, either in the short or long term.

Payback Analysis: The payback analysis method is a popular way to figure out whether or not an investment will pay for itself. As previously said, system development requires more expenses in the early stages, as well as a long time for the benefits to outweigh the costs. The majority of the expenditures in the early stages of system development are spent on analysis, design, and implementation. The cost is required after installation to cover the running expenditures that must be recouped. Payback analysis analyses how long it will take for the benefits to outweigh the expenditures. This is referred to as the payback phase. The payback period is the amount of time it takes for cumulative benefits to outnumber acquired and ongoing expenditures.

Return-on-Investment Analysis (ROI)*:* The ROI (Return-on-Investment) analysis approach evaluates the long-term profitability of different alternatives. The return on investment (ROI) of a solution is a percentage rate that evaluates the link between the quantities received by the firm and the amount invested.

Lifetime ROI = *(Estimated lifetime benefits – Estimated lifetime costs)/Estimated lifetime costs*

Annual ROI = *lifetime ROI/lifetime of the system*

Net Present Value: The term "net present value" refers to a method for comparing the annual costs and advantages of various options. For each year of the system's life, we specify the costs and benefits. The current worth of any object at any point in the future is known as its present value.

Present Value (PV_n) = $1/(1 + \mathbf{i})^{\mathbf{n}}$

where **n** is the number of years and **i** is the discount rate

(c) ***Operationally Feasibility***

It's a metric for determining how effectively a solution satisfies the system's criteria. Operational feasibility is a measure of how effectively a solution fits the system requirements to address issues and capitalize on opportunities discovered throughout the scope definition and analysis process. It assesses how well it meets the system requirements defined throughout the requirements analysis process. It's also concerned with the existing problem and the expense of the proposed remedy, as well as if the problem exists or not. It also answers the following questions:

1. How successfully the proposed system addresses issues and exploits opportunities identified during the scope definition and problem analysis phases.
2. The proposed system's ability to meet the system requirements was defined during the requirements analysis.

Table 7.2 Feasibility analysis matrix

	Weightage (%)	Candidate 1	Candidate 2	Candidate 3
Description		*Technology 1*	*Technology 2*	*Technology 3*
Operational feasibility	30	60	70	80
Technical feasibility	30	70	60	70
Economic feasibility	40	80	90	60
Weighted score	100	73	75	69

7.4.6 Feasibility Analysis Matrix

At the third checkpoint of the feasibility study, the group discovers all of the potential system solutions and then analyses each of those options. To identify which of the applicants is the most qualified for the position, we evaluate each of them side-by-side. There are two methods for making a comparison and a recommendation. The candidate system matrix and the feasibility analysis matrix are the two matrices. The potential systems are analysed and ranked in the feasibility analysis matrix. It contains the same columns as the candidate system matrix, plus one more called a ranking column. While all of the feasibility limitations are important, they pale in comparison to technical and operational feasibility. As a result, we may create a weighting system with a modest ranking for constructing the feasibility analysis matrix.

Weightage scheme for feasibility analysis matrix

Key issues	Weightage (%)
Operational	x
Technical	y
Economic	z

Table 7.2 depicts an example of a feasibility study matrix that was created during the project's solar panel selection. We can see in the table that there are three different solar panel technology with different properties. Some are good in terms of operational feasibility, while others are good in terms of economic feasibility. So, to make a viable judgement about whether to embrace any of them, we created an aggregate weighted score for each technology, and based on that score, that technology was chosen. Table 7.2 organizes all the qualities into a logical order.

7.4.7 Feasibility Analysis with HOMER Simulations

HOMER is the gold standard when it comes to microgrid software because it was developed after decades of listening to the needs of customers from all over the world. These customers have expertise in developing and installing microgrids and distributed power systems that can include a mix of fossil-fuelled production, storage, and renewable energy sources. HOMER was built to meet these customers' demands (either through a local generator or a power grid). The HOMER (Hybrid Optimization of Multiple Energy Resources) models simplify the process of constructing hybrid renewable microgrids, regardless of whether they are located off the main grid or are linked to one. The optimization and sensitivity analysis methods that are included in HOMER make it possible for us to evaluate the economic and technical viability of a wide variety of technological options while considering differences in the costs of the technologies, the amount of electric load, and the availability of energy resources. The National Renewable Energy Laboratory was responsible for the development of the HOMER system as part of their village power initiative. Currently, the HOMER system is licensed to HOMER Energy. HOMER, or the micropower optimization model, makes analysing off-grid and grid-connected power systems for a range of applications easier. When developing a power system, there are several aspects of the system's setup that need to be considered, including the following: What are the different parts of the system that need to be included in the design? How many of each component should you use, as well as their dimensions, in the finished product? These issues are made more difficult by the vast number of technical choices that are accessible, as well as the broad range of prices associated with various technologies and the availability of various energy resources. The optimization and sensitivity analysis methods that are included in HOMER make the process of analysing the myriad of possible system configurations a lot less difficult (Fig. 7.6).

The selection of the appropriate system components and resources to include in your model is the first step in commencing an analysis with the HOMER version 2.68(beta) software. Figure 7.7 is a snapshot of the HOMER software's main interface. PV components must first be added to the model under the equipment to Consider tab, which is situated in the upper right corner of Fig. 7.8, to develop a model that can estimate solar PV power potential. The grid was then updated with the specified PV component. A snapshot of the equipment options accessible in the HOMER tool is shown in Fig. 7.7. For the project depicted, the predicted simulation in HOMER software will do a probability analysis.

The next stage is to add an inverter to the model, which is referred to as a converter in HOMER. Because there are so many different types of inverters on the market, deciding which one to use was a challenge.

This was particularly difficult because the PV type was so important. To overcome this problem, the default parameters of the PV option were altered to output AC directly, rather than configuring the PV output as direct current (DC) and requiring an inverter to convert it to alternating current (AC). Figure 7.8 shows how a traditional solar PV system is linked to the grid via an inverter using the HOMER option. In the next part, the PV parameters that were entered into the model and their values are reported, and Fig. 7.9 shows a snapshot of these PV parameters.

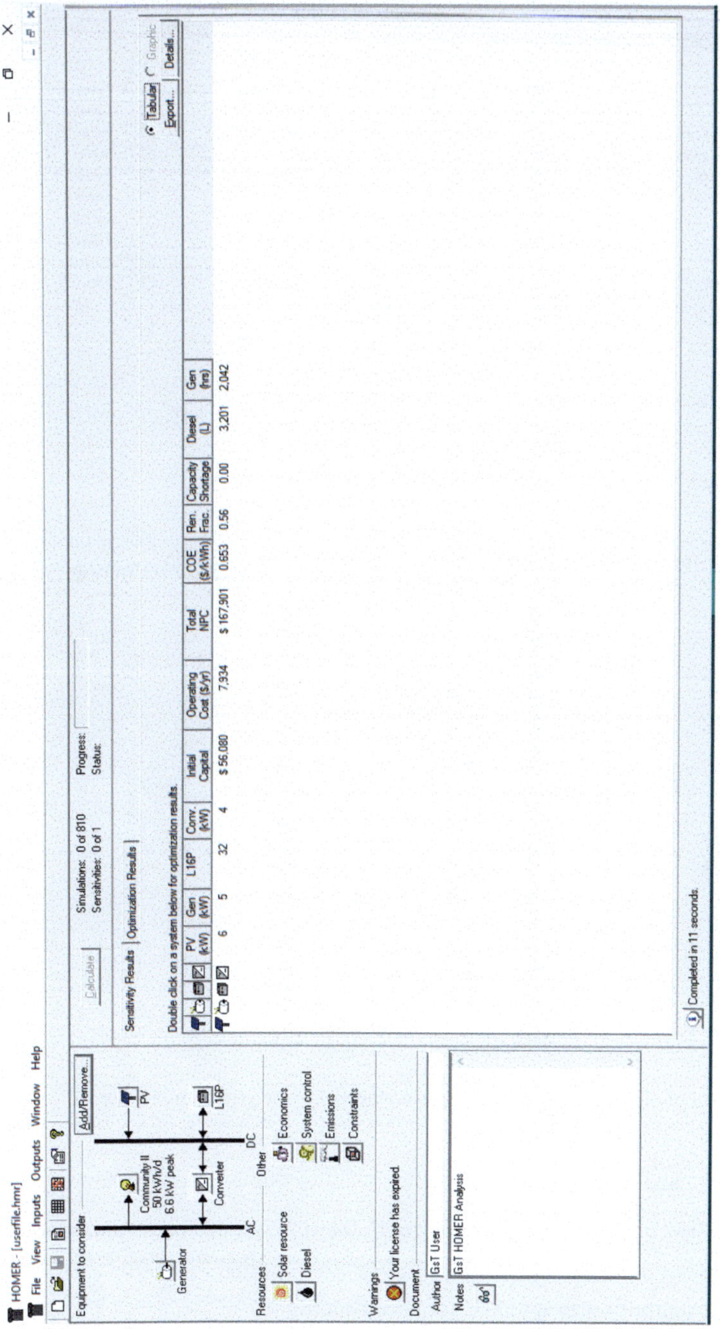

Fig. 7.6 Illustration for the main window of HOMER software

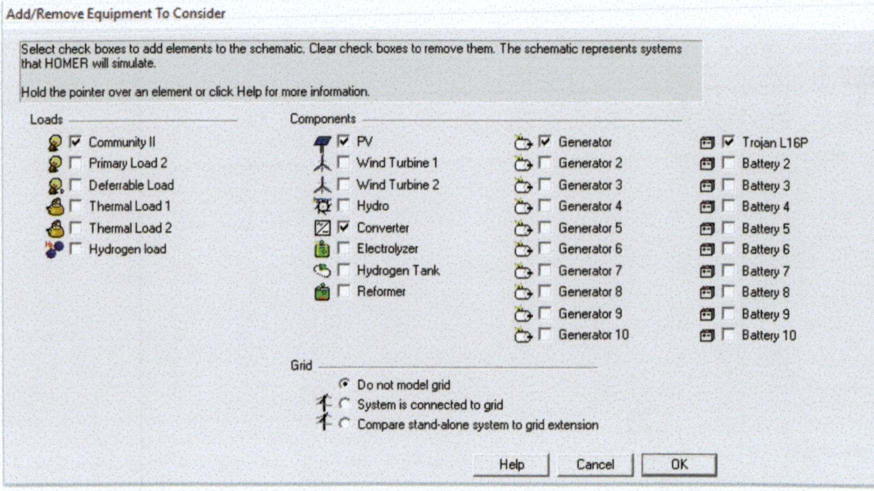

Fig. 7.7 HOMER add/remove equipment to consider the option

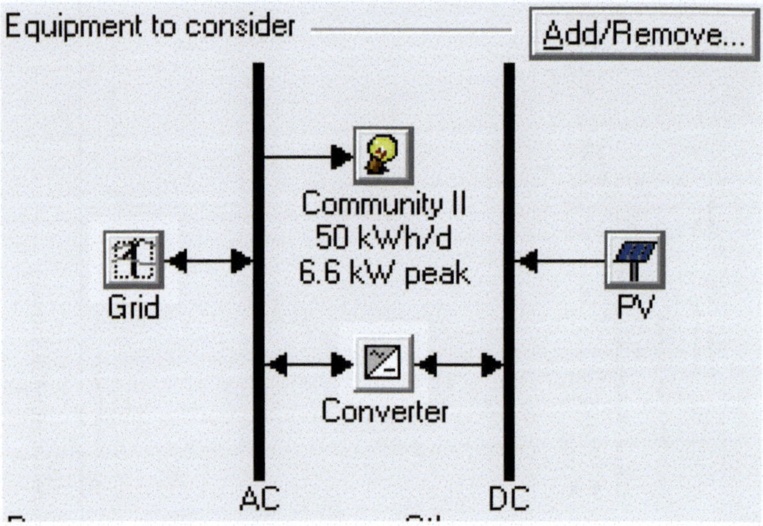

Fig. 7.8 Illustration of conventional PV system connected to the grid with inverter

The PV inputs used are:

Slope—The slope of the PV panels was set to 0° to maximize roof space and panel installation.

Azimuth—The azimuth was set to 0° for south-facing.

Tracking system—No tracking system was used for the modelling.

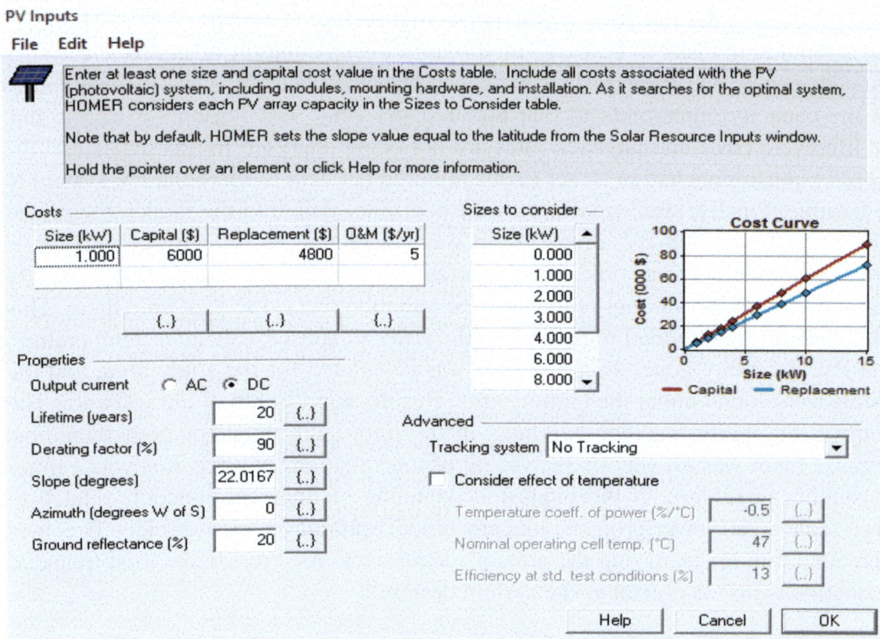

Fig. 7.9 Input window for PV input to be used for analysis

Ground reflectance—This was set to a value of 20% to account for the third component of radiation due to albedo from the ground. This 20% value is a valid approximation, as it does not particularly influence the results in any significant amount.

Derating factor—A derating factor is a scaling factor that accounts for losses in the system due to dust, wiring or anything not ideal. The value of 80% is the default estimate value used by HOMER.

PV array size—This refers to the size in kW of the desired PV array.

The next task after setting up the PV inputs was to set up the solar resource data. The Solar Resource Inputs window requires the latitude, longitude, time zone, and baseline data. The latitude and longitude input affects the curve of the graph of the baseline solar resource data. In particular, the latitude of the site affects the breakdown between the direct and diffuse radiation that affects the timing and the heights of the peaks in the solar resource. The southern latitudes receive more solar resources, and these values gradually decline throughout the day, whereas northern latitudes receive less solar resources and experience larger differences in peaks throughout the day. The longitude, on the contrary, affects the timing of the power output; thus, the location of the site has a very significant impact on the PV power output of HOMER. HOMER recognizes the baseline data in one of three formats: monthly average worldwide solar radiation on the horizontal surface, monthly average clearness index, or hourly average global solar radiation on the horizontal surface (Lambert, 2006). There are a few remaining tasks to be performed before the calculation of hourly electricity can commence, and these tasks were to set the electrical grid rates to zero and to set the capacity constraints to one hundred per cent. The reason for this is that the life-cycle costs and payback rates are not being analysed in this thesis and the capacity constraints had to be set to one hundred per cent to force the PV system to be feasible. Finally, after inputting all the necessary data into the model, the power production from each PV array system was calculated. In HOMER, photovoltaic arrays are modelled as an object that generates electricity in a manner that is exactly proportionate to the amount of global solar radiation that is incident onto them. This relationship is maintained regardless of the array's exposed voltage or temperature. The PV electricity generation potential was calculated for the study area, and the results were found under the Optimization Results window tab of the software. The analysis was performed for buildings in the pilot study area for both scenarios. Because there was no way to analyse all of the situations at once, this was a time-consuming procedure. A thorough understanding of financial concerns related to large-scale solar power projects, like any other capital investment project, becomes a precondition for justifying the project's legitimacy. As a result, the total financial evaluation is just as crucial as the system design.

7.4.8 SWOT Analysis~Vitality and Feasibility

The SWOT analysis is the first and foremost tool for considering the viability and prospects of any research project or concept. The benchmark for evaluating the feasibility of research efforts in the international scientific arena is an important aspect of the SWOT analysis (Table7.3).

The connections of the four dimensions Strength (**S**), Weakness (**W**), Opportunity (**O**), and Threat (**T**) served as the core for the analysis of future strategies:

1. **Strength and Opportunities (SO)**: To use strengths to take advantage of these opportunities.
2. **Strength and Threat (ST)**: To take advantage of strengths to avoid real and potential threats.
3. **Weakness and Opportunities (WO)**: To use opportunities to overcome the weaknesses experienced in the research.
4. **Weakness and Threats (WT)**: To minimize your weakness and avoid threats.

7.5 Results and Discussions

Like what the internet accomplished for computers, electrical utility grids are undergoing a major paradigm change in their usage of information technology and communication technologies. Although it is impossible to predict how a smart grid would

Table 7.3 SWOT analysis matrix for vitality and feasibility of current research

	Strength	Weaknesses
Opportunities	**Strength**: Instant access to research problems for global development **Opportunity**: Growing scientific and societal interest in geo-awareness **Strategy**: Organize one umbrella research theme where all disciplines can participate	**Weakness**: Poor visibility of research themes in the country **Opportunity**: Can contribute to the international development agenda **Strategy**: Take the initiative to join forces for expertise for the global development
Threats	**Strength**: Directly applicable full spectrum Geo-Information science **Threat**: Geo-ICT is not valued by policymakers **Strategy**: Promote our science within the country and abroad	**Weakness**: A tension exists between the capacity mission and excellence in fundamental research **Threat**: Funding in future may disappear **Strategy**: Define relevant criteria for quality in capacity building and research and attract more diverse(private) funding

appear in 20 years, existing technologies may be implemented, and efforts taken to prepare the utility grid for future innovation. Given the likelihood that renewable energy penetration levels will continue to rise, a rethinking of the current energy balance paradigm may be necessary. Fortunately, a functioning smart grid has the potential to alleviate some of the problems that large levels of renewable energy output can cause. A smart grid takes use of the potential for conventional operations to be improved via the use of connectivity and information. Distribution System Automation (DSA), Selective Load Control (SLC), and Managing Distribution are some of the features of the smart grid that help with distribution system control and operation. While renewable energy cannot always be controlled traditionally, its behaviour can be forecasted to increase system efficiency. As a result, we can see that renewable energy has a distinct advantage over smart solar grids (SSG), as energy scarcity has become a worldwide issue, and developing countries require an efficient energy system to reduce losses and maximize the use of generated electricity. The only correct answer is the smart grid system. The energy picture is presented in terms of overall energy demand and supply, as well as its context. The opportunities for renewable energy have been addressed in this article. To use the HOMER tool, provide the following economic inputs (as indicated in Fig. 7.10).

Figure 7.11 shows the cash flow summary (economic feasibility) of a solar photovoltaic project with a 6 kW PV panel, a 5 kW generator, a 4 kW inverter, and a 4 kW rectifier, broken down by component. We will get the illustrated outputting of the same cash flow summary when these component cash flows are simulated with a given set of input values. We have considered the capital cost, replacement cost, operational and maintenance cost, fuel cost, salvage cost, and finally the total cost of all these things.

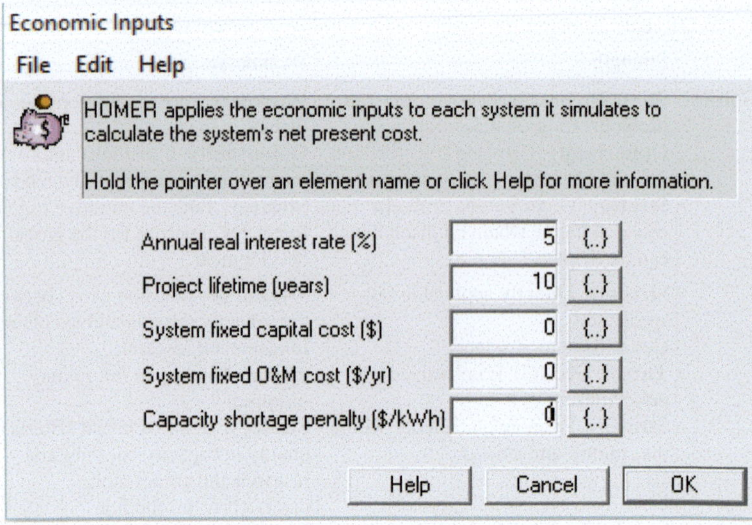

Fig. 7.10 Input window for economic inputs

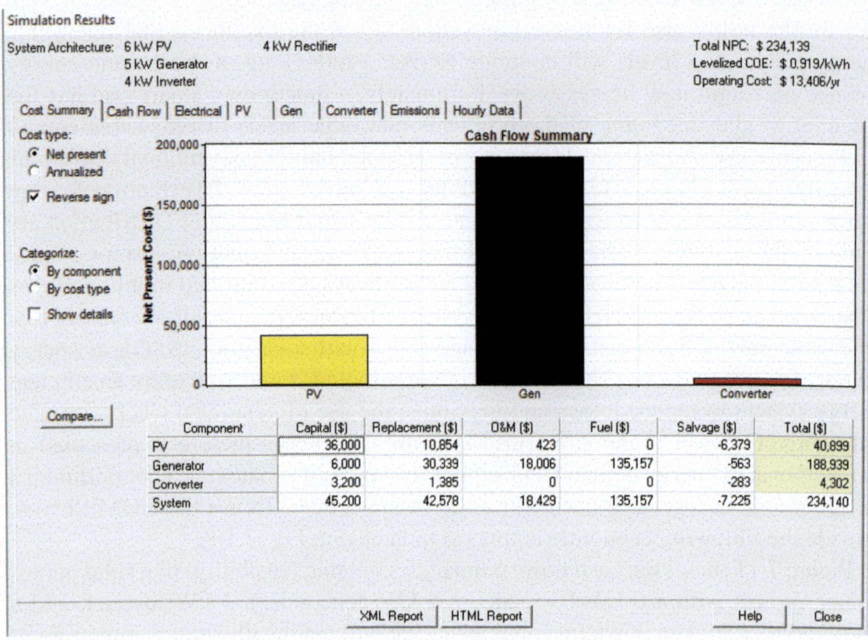

Fig. 7.11 Output window generated from provided economic inputs

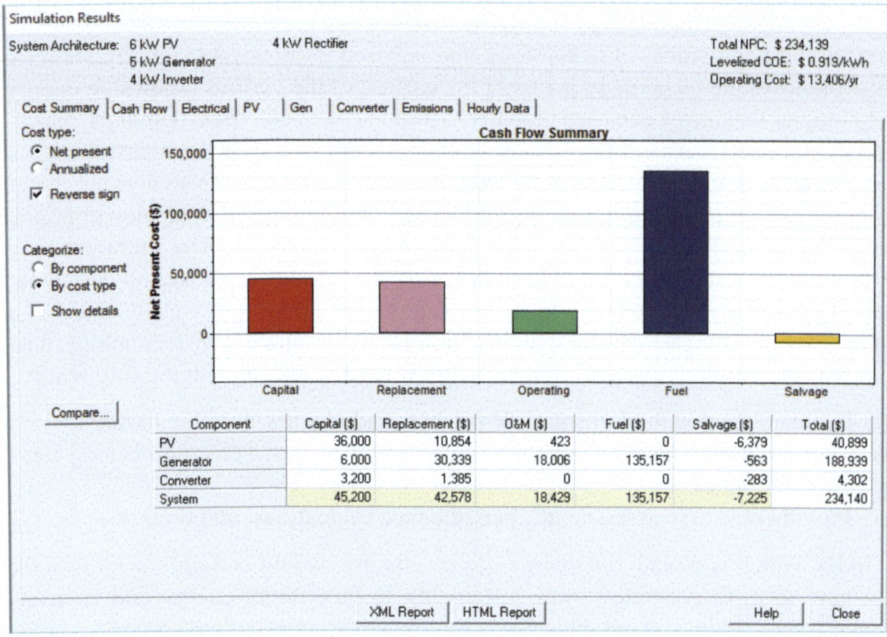

Fig. 7.12 Simulated results generated from provided economic inputs

Figure 7.12 depicts the cash flow summary cost type of a solar photovoltaic project with a 6 kW PV panel, a 5 kW generator, a 4 kW inverter, and a 4 kW rectifier. We will get the illustrated outputting of the same cash flow summary when these component cash flows are simulated with a given set of input values. We have considered the capital cost, replacement cost, operational & maintenance cost, fuel cost, salvage cost, and finally the total cost of all these things. It also calculates the total NPC, the levelized cost of energy (COE), and the operational cost. The simulation run results are displayed in the above figure under several tabs such as cash summary (as seen in Fig. 7.12), cash flow, electricity, PV, generator, and converter, as well as hourly generated data.

7.6 Summary

As a result of the rapid use of fossil fuel sources, it is possible that the world could run out of energy supplies in the not-too-distant future. This should be a reason for concern in developing nations since their economies are so heavily dependent on the use of energy. Renewable energy sources must be used with the maximum possible conversion efficiency to keep up with the ever-increasing demand for energy. In addition, since international economic and political conditions tend to make nations more dependent on their energy supply, there has been a growing interest in the development and deployment of technology that is based on renewable energy sources. Solar photovoltaic technology is one of the earliest of the various renewable energy technologies that have been used globally to provide basic electrical demands, particularly in isolated places. This is because solar energy is one of the oldest forms of energy that can be harvested directly from the sun.

This section presents the findings of research that conducted a techno-economic feasibility analysis of producing solar photovoltaic electricity. The literature may be broken down into three basic categories: method designs, the techno-economic viability of solar photovoltaic power production, and performance evaluations of a variety of different systems. Our expertise in solar photovoltaic (PV) technology, data analysis, and building and grid infrastructure is used to accomplish the following:

(a) Evaluate the potential of solar PV projects and identify possible hazards.
(b) Plan and design installations for best performance, based on credible long-term yield estimates.
(c) Provide objective assessments, performance evaluations, and reports.

India, which is already an energy-scarce country, would certainly need to look into new ways to generate power sustainably to face future energy and environmental issues. India's geographical position makes it an excellent choice for solar energy harvesting. As a result, solar PV technology can fulfil India's future energy needs while also addressing environmental concerns. Solar PV electricity has been identified as a potential answer to future energy concerns in the current study.

Solar PV-generated power will be used to satisfy energy demand during the day in the planned smart solar grid energy network. The extrasolar photovoltaics generated will be stored in a variety of storages that may be used during midnight hours and extended cloudy circumstances. For all the suggested technologies, current and anticipated cost scenarios have been supplied to assess the economic feasibility of the energy network under consideration. PV power is predicted to be more cost-effective than fossil-fuel electricity based on future trends.

Most of the current researchers has focused on renewable energy resources to make the country's energy independent through locally designed and generated solar power. The suggested system design based on smart grid methodology described the efficient system of PV system to be the then primary power source. Solar panels will be put on the roofs of power customers. The simulation reveals that feasibility is a cost that is both competitive and feasible. HOMER simulation and assessment were used to do an economic study of a planned renewable energy system. This research shows how solar radiation data might be used to propose a future solar-powered electrical network.

References

Aleksandrowicz L, Green R, Joy EJM, Harris F, Hillier J, Vetter SH, Smith P, Kulkarni B, Dangour AD, Haines A (2019) Environmental impacts of dietary shifts in India: a modelling study using nationally-representative data. Environ Int 126:207–215. https://doi.org/10.1016/j.envint.2019.02.004

Alizadeh MI, Parsa Moghaddam M, Amjady N, Siano P, Sheikh-El-Eslami MK (2016) Flexibility in future power systems with high renewable penetration: a review. Renew Sustain Energy Rev 57:1186–1193. https://doi.org/10.1016/j.rser.2015.12.200

Borfecchia F, Caiaffa E, Pollino M, Cecco LD (2014) Remote Sensing and GIS in planning photovoltaic potential of urban areas. Eur J Remote Sens 47(195):195–216. https://doi.org/10.5721/EuJRS20144713

Cooper CM, Troutman JP, Awal R, HabibiH, Fares A (2022) Climate change-induced variations in blue and green water usage in U.S. urban agriculture. J Cleaner Prod 348. https://doi.org/10.1016/j.jclepro.2022.131326

Dovì VG, Friedler F, Huisingh D, Klemeš JJ (2009) Cleaner energy for sustainable future. J Clean Prod 17(10):889–895. https://doi.org/10.1016/j.jclepro.2009.02.001

Farhangian Marandi O, Ameri M, Adelshahian B (2018) The experimental investigation of a hybrid photovoltaic-thermoelectric power generator solar cavity-receiver. Solar Energy 161, pp 38–46. https://doi.org/10.1016/j.solener.2017.12.039

Gils HC, Scholz Y, Pregger T, Luca de Tena D, Heide D (2017) Integrated modelling of variable renewable energy-based power supply in Europe. Energy 123:173–188. https://doi.org/10.1016/j.energy.2017.01.115

Islam MS (2018) A techno-economic feasibility analysis of hybrid renewable energy supply options for a grid-connected large office building in Southeastern part of France. Sustain Cities Society 38:492–508. https://doi.org/10.1016/j.scs.2018.01.022

Lewis NS, Crabtree G, Nozik AJ, Wasielewski MR, Alivisatos P (2005) Basic research needs for solar energy utilization. Basic energy sciences workshop on solar energy utilization, vol 276. http://www.sc.doe.gov/bes/reports/files/SEU_rpt.pdf

Li C, Xu C, Li X (2020) A multi-criteria decision-making framework for site selection of distributed PV power stations along high-speed railway. J Cleaner Prod 277. https://doi.org/10.1016/j.jclepro.2020.124086

Nayan F, Islam A, Mahmud S (2013) Feasibility Study of Smart Grid in Bangladesh. 2013(July):1531–1535. https://doi.org/10.4236/epe.2013.54B290

Rybach L (2014) Geothermal power growth 1995–2013—a comparison with other renewables. Energies 7(8):4802–4812. https://doi.org/10.3390/en7084802

Tsoutsos T, Frantzeskaki N, Gekas V (2005) Environmental impacts from the solar energy technologies. Energy Policy 33(3):289–296. https://doi.org/10.1016/S0301-4215(03)00241-6

Zhang C, Wang Q, Zeng S, Baležentis T, Štreimikienė D, Ališauskaitė-Šeškienė I, Chen X (2019) Probabilistic multi-criteria assessment of renewable micro-generation technologies in households. J Clean Prod 212:582–592. https://doi.org/10.1016/j.jclepro.2018.12.051

Chapter 8
Synthesis and Transitions in Energy Systems Policies

Abstract Energy is a key variable in today's world, and it is important for any society's socioeconomic well-being and stability. Concerns about the insecurity and fragility of fossil fuel supply, as well as the numerous and varied ways in which they harm our environment, have prompted a gradual but steady shift in the existing energy balance. Increased reliance on renewable energy sources is required for long-term growth. Solar energy, which has been used since the dawn of time, is now gaining traction as the "future" fuel, with enormous potential to supply the world's energy demands at any given time. Understanding the sun's energy pattern on Earth is needed to utilize it. Solar resources change geographically and temporally across Earth's surface. Solar applications start with a resource evaluation. Designing and modelling solar projects need solar data. Solar radiation measurement is expensive. Due to a dearth of ready-to-use solar radiation data, radiation must be estimated. Solar applications need diffuse and direct components to calculate radiation received by sloped surfaces. This emphasizes the need of calculating any of them from globally measured radiation to get the third component as the difference. These gadgets come in many forms. This work aimed to create realistic, basic diffuse radiation models. Adding meteorological elements to clearness index-diffuse ratio models improved the estimate. Renewable energy sources create power with a low input of carbon dioxide (CO_2) or other greenhouse gases (GHG) to the environment, as well as no damaging discharge on water or land. Solar, wind, hydro, and biomass are renewable energy sources with huge potential.

8.1 Synthesis of the Research

8.1.1 Originality of Contribution

To this day, topographical characteristics continue to be the single most important element in determining how well solar insolation can be described and, as a result, how accurately solar radiation can be estimated. However, despite this fact, it is still permissible to investigate other significant factors to improve the estimate accuracy. People's attention has been sparked in the past, individually, by the amount of time

D. Kumar, *Renewable Energy Scenarios in Future Indian Smart Cities*, Advances in Geographical and Environmental Sciences, https://doi.org/10.1007/978-981-19-8456-3_8

that the sun is out, the cloud cover, and the air mass. On the other hand, the current research effort was designed to investigate and compare the effects of factors such as air conditions, altitudes, surface orientation, and topographic effects on global solar estimations on a large scale. This was accomplished by investigating a global scale. This strategy has a practical foundation since there are fewer meteorological stations documenting sunlight duration, cloud cover, temperature, and so on. Additionally, fewer stations are measuring solar radiation. Applications that use solar energy, like many others, need data that is easily accessible, site-specific, and long-term. On the other hand, quantifiable data sets on the sun's radiation are sometimes lacking. The addition of data such as latitude, day of the year (DOY), local terrain configurations, and climatological factors such as a sunny fraction (SF) and cloud cover (CC), among others, promotes the research's practicability for larger applications. Within the framework of the study that has been provided, the following subsections will highlight original contributions as follows:

Database: According to the author's best knowledge, there has never been single research undertaken in the history of solar radiation modelling that employs such a broad variety of locales. The latitudes used in this study range from 15.50° N to 17.50° N, and the longitudes range from 76° E to 78° E.

Approach: The influence of sunlight percentage, cloud cover and air mass, or even just the two of them together with the latitude, day of the year (DOY), and local topography configurations were not included in any of the models that were identified in the prior research that was published in the scientific literature.

Temporal scale: Most of the earlier studies that looked at global radiation took into consideration yearly, decadal, monthly, or even, in some instances, daily averages of radiation levels. In addition, no research was carried out over such a long time as the years 2004–2014 combined. The availability of multi-parameter models for the estimate of local season values is quite limited. In contrast to the models that are now being presented, the vast majority of the daily radiation models that were established in the past were very site-specific.

Methodology: The models that were constructed here are inevitably distinct from the many-parameter models that are accessible in the literature. These models are typically described by an empirical equation of the first degree, and subsequent parameters are organized sequentially. The present progression modelling makes use of additional descriptors as sub-functions to provide better explanations of the research being conducted. These descriptors include latitude, day of the year, local terrain configurations, and topography. The process of optimization is carried out on every level, from the different degrees that may be achieved within a single parameter qualification to the inter-comparison that takes place between the three variables that are offered.

Evaluation: Comparing monthly averages is used to assess a good number of the hourly radiation models that have been developed in the past. When hourly measured and calculated data are compared, the substantially increased dispersion and variation that emerges from this comparison are hidden using such an average. Because of this, it successfully improves model correctness in contrast to assessment that is done hour by hour. The data that is produced hourly is what is utilized to analyse all the statistical indicators in this research. The great majority of daily radiation models that can be found in the scientific literature have been verified or evaluated making use of the very same data that was used to develop them. In this research, daily models are compared against independent data, which either come from a different region or a different time. Both factors are taken into consideration.

Statistical approach: Precision Score (PS), a new comparative index that values the credits of all currently employed statistical indicators to provide a common and unique ground of inter-comparison across all models, was recently introduced. PS does this by valuing the credits of all currently employed statistical indicators. Despite an exhaustive search of the available research, there was not a single paper that included such a precise statistical assessment for several models of solar insolation. Most of them made use of the well-known MBD and RMSD, while just a handful of them used the R^2 strategy. Nevertheless, the Precision Score was also used in this research project (PS).

8.1.2 Observations from Research

The present research work reported the following experimental results:

1. Solar Energy Modelling, Data processing, and Solar Map creation techniques.
2. Incident solar radiation estimated average daily solar exposure and potential sites for PV energy harvest.
3. Hourly electricity and cumulative hourly PV electricity estimates in kW.
4. Rooftop area calculations and feasible areas on the rooftop for PV installations.
5. Approaches to compute the Photovoltaic (PV) energy potential for the buildings.
6. Solar insolation variability analysis on Spatio-Temporal and Hyper-Temporal scale.
7. Spatial variation of surface solar energy fluxes for land-use changes.
8. Feasibility Analysis for Smart Solar Grids and Demand–Supply Analysis for solar energy.
9. Study on Renewable Energy Potential for Green Networks.

Within the same data set, it was found that one model based on location, DOY, and CF parameters performs much better than the other model that has been constructed. In other words, including local topography characteristics in models improves the accuracy of estimates, particularly when the models are used to fill in gaps in the actual data set, such as estimating diffuse irradiance in areas where it was not recorded due to station shutdown or other operational concerns. This is especially true when the models are used to fill in gaps in the actual data set. These models, on the other hand, are site-specific and, in some cases, time specific as well. This is because meteorological standards have been poorly maintained and there are unavoidable and systematic inaccuracies connected with such factors.

The primary takeaways from this study are that qualitative research on the suggested parameter using more descriptors is required since they are the findings that emerged from the research. During the bulk of the study, it was established that every model was location dependent. The provided models, although site-specific to a significant degree, have the same parameter combination for the whole area and have a propensity for their coefficients to be generalized across locations with meteorological and climatological features that are comparable to one another. The model types were found to be consistent when compared to neighbouring stations or other locations with circumstances that were comparable. As a result, an overall best model for the specified area was suggested. The previous generalization used a methodology that may be described as quite optimistic. This must be reviewed twice, and the coefficients may need to be adjusted slightly to accommodate the requirements of a variety of geographic areas. It was determined that this activity, which was yet another extensive one, did not fall within the parameters of the present project.

When it comes to applications, diffuse radiation models need to be site independent as much as possible. This is a goal that is very difficult to achieve since diffuse irradiance varies depending on various characteristics that are specific to a particular location and are hard to generalize. Nevertheless, there is always the option of standardizing the models for climates and turbidity levels that are comparable to one another, as well as other factors. Following an exhaustive investigation, it was found that, next to latitude, day of the year (DOY) is the factor that has the greatest influence on the production of global solar radiation, and that the use of models has enhanced accuracy. In contrast to the other two criteria, such as latitude and topographic characteristics, it was found that using DOY results in a more accurate estimation, particularly at the higher end of the range of measured global solar radiation. This was the case even though there was no sign of a subjugation trend. According to the validation approach, it is unnecessary to consider additional factors such as CF if daily diffuse radiation is measured at a different location. The degree of connection that exists between the other two elements is the only thing that differentiates all the suggested regional models that incorporate just latitude as an extra component from one another. Because of this, we are confident in putting up this model as a potential regional model.

8.2 Research Outcomes

Chapter* 1: *Concept of Sustainable Energy System for Smart Cities

This chapter introduces the concept of sustainable energy system for smart cities of any study area, a brief research methodology, and the research's interest.

Chapter* 2: *Overview to renewable energy systems

This chapter introduces the problem, the study's reason, the objective, the experimental study area, a brief research methodology, and the research's organization.

Chapter* 3: *Advancement in Renewable Energy Scenarios

This chapter covered a thorough literature analysis of solar energy and photovoltaic (PV) systems, as well as in-depth examinations of the motivations for using renewable energy sources. Solar energy, challenges, and factors affecting rooftop PV systems in the urban context were examined, as well as a thorough examination of the software tools used to estimate solar energy potentials. This gave a thorough examination of the solar software tools, as well as criteria to consider when choosing one of the solar tools for any type of solar PV generation potential study. Based on the study's aims, it outlined the significant research projects currently underway in the sector. The chapter also gave a comprehensive list of the software that might be used to do the research. A comparative list in tabular format was produced to summarize the software's strengths, weaknesses, advantages, and disadvantages in use. Finally, it outlined the chapter's significant contributions to future research. This chapter identified known research gaps that must be considered when developing any modelling tools. Aside from that, it described the strategies and methodologies that might be used to conduct evaluation studies on the generated model.

Chapter* 4: *Conventional Methods of Energy System Evaluation

Because solar insolation time series data are increasingly being used for solar resource mapping and monitoring, developing a simple technique for modelling solar energy with advanced priority is a must. To represent the solar potentials of any place, a more accurate and realistic solar map demonstration is required. This chapter discussed existing and newly developed methodologies for modelling solar energy using geospatial technology, considering various components such as a global, direct, and diffuse beam of terrestrial solar radiation, as well as their effects on global solar radiation. It attempted to quantify the available worldwide sunlight by taking into consideration numerous elements including latitude, day of the year (DOY), local topography layout, and atmospheric transmittances. To produce better results, the model was evaluated for sensitivity to each of the described factors. After numerous experiments, an approach known as the spatial solar radiation model, or S^2R model, was developed. As an input to the model, satellite-derived datasets such as digital elevation model (DEM) are required. The model will compute global solar radiation over one year based on our demands, namely area-wise solar insolation or point as solar insolation. It also allows users to input values for limitations or controlling parameters such as latitude, day of the year (DOY), and desired output image resolutions.

The model's sensitivity was tested using characteristics such as latitude, day of the year (DOY), local topography layout, and atmospheric transmittances. Based on the results of this sensitivity exercise, a final model was created that can withstand all minor perturbations caused by problems with spatial extend/resolution, atmospheric transmittances, and site configurations. The model was also given the ability to endure minor data set disturbances caused by elevation, surface orientations, shadows produced by adjacent topography, and daily and seasonal fluctuations in solar angles, as well as atmospheric attenuation.

However, the produced model's results adjudge the current solar power plant locations within the Kalaburagi district, and these have been discovered by the model's results. Predictive models must therefore deliver trustworthy findings to fully leverage the benefits of solar systems. The precision of global solar radiation calculated as a raster image using satellite-based solar radiation (S^2R) models is shown to be dependent on the following factors:

1. Simulation accuracy of cloud transmittance generated from satellite data.
2. Quality of input parameters characterizing actual conditions of the atmosphere, such as aerosols and water vapour.

Because there was a reasonable chance of error in the datasets, it was necessary to eliminate uncertainty in the datasets generated. It must be assessed using some well-established methodologies. However, traditional approaches for evaluating precision datasets, such as the coefficient of determination (R^2), mean bias deviation (MBD), root mean square deviation (RMSD), skewness, and kurtosis, fail due to the increased ambiguity and non-standard or generic data formats provided by the model. As a result, evaluating the data sets necessitated the use of a conventional or novel procedure. In this regard, we devised a novel method for evaluating modelled datasets using other satellite-derived information. This method employed the MBD and RMSD values to rank the inaccuracy in modelled datasets, which may be used as a standard method to evaluate the data sets (as mentioned in Sect. 4.4.5 of this thesis).

Similarly, processing the raw data using the right data processing technique was necessary to acquire the desired outcomes from the intermediate values. The data processing section went over the various data processing techniques that were used. It also established a mechanism for processing the modelled data (as illustrated in picture 4.10 in Sect. 4.4.6). The viability of various data processing approaches has also been discussed.

Finally, the chapter focuses on how to create sample solar panels. The formulated method of solar map production (as illustrated in Fig. 4.11 in Sect. 4.4.7 of this thesis) has been elaborated in microscopic detail. These solar maps' utility has been briefly discussed. To familiarize the audience with the developed approach for solar map production, sample solar maps were displayed (Fig. 4.12 in Sect. 4.5 of Chapter 4). The chapter ends with illustrated solar maps of the study area, which allow for visual interpretations of accessible global solar resources.

Chapter* 5: *Quantifying Amount of Energy Harvest

The natural and semi-natural landscape cover is varied, with progressive spatial shifts. Because solar insolation/flux heterogeneity is a non-static phenomenon, it is influenced by temporal changes and landscape cover. This chapter aims to discover and measure solar insolation for photovoltaic yield estimations, as well as suitable solar energy harvesting sites. Modern satellite-based databases have a high spatial and temporal resolution, making them the go-to source for solar energy applications like solar power plant operation and management. The consistency of daily global radiation estimations in complicated topographical areas was identified in this section.

A digital elevation model was used to create high-resolution solar radiation maps under clear skies. We also looked at the sensitivity of solar radiation maps in the test area for numerous DEM resolutions and different topography characteristics. The geography of the research area may have an essential influence on the model's reliability to generate precise solar insolation maps under clear-sky conditions (or severe shadow-castings), according to the findings. The global solar insolation data were initially quantified on an area-by-area and point-by-point basis. However, before we can quantify solar insolation, we will need reliable software/tools to do so. As a result, software selection criteria were created based on the requirements of our work to select the best tool. After that, solar insolation numbers were calculated on an area-by-area and point-by-point basis. The method/techniques used in these computations (area-wise and pointwise) are described in Sects. 5.4.5 and 5.4.6 of Chapter 5.

In addition, the work created the thumb rule for calculating the minimum area required to generate one kilowatt of photovoltaic power. According to Sect. 5.4.7 of Chapter 5, installing a photovoltaic system that generates 1kWh of electricity requires a minimum of 3 m^2 of space. Aside from that, photovoltaic yield predictions for the shown pilot study area were performed under several configurations. It also included strategies for extracting rooftop spaces as well as computer-usable rooftop sections.

The chapter's findings detail the area-wise and pointwise solar insolation maps in great detail. It also accounts for and displays the pilot area's power generation calculations. For further clarity, the acquired results were presented in tabular form for the various scenarios. With roughly 5.09–5.48 $kWh/m^2/day$ annual global sun insolation, the study highlighted the southern part of the Kalaburagi area as the most potential region with good solar insolation, while the rest of the locations were likewise found to have a high potential for global solar radiation.

However, the current chapter supplied solar insolation quantity estimates, which were then used to calculate potential photovoltaic (PV) yield estimates. Although great attention was taken to quantifying solar insolation, data from in-situ stations, as well as the inclusion of climatographic elements such as sunshine hours, fog, altitude, and other factors, will have a direct or indirect effect on solar energy and are likely to produce improved findings.

***Chapter* 6: *Variability of Energy Resources at Diverse Scale*.**

Solar energy is best used in city centres and other urban areas. It is anticipated that large-scale thermal and photovoltaic (PV) applications will be deployed in metropolitan areas in the future. In urban areas, the value of global and regional solar resource databases with spatial resolutions ranging from 100 m to one kilometre is restricted. Solar energy, in contrast to more traditional forms of electrical power production (such as those based on fossil fuels or nuclear power), is unreliable. The output of a solar power plant is affected not only by the climate but also by the passing of the days and the changing of the seasons. The present research examines satellite-derived information on a variety of geographical and temporal dimensions to give reasoning and evidence about the phenomenon of intermittent solar energy. In the records kept at weather stations, data on solar irradiation might be difficult to locate at times. Some approaches may be more successful than others in addressing these problems. In this chapter, we illustrated how variability statistics may be used to get a better understanding of the breadth of measurements that are necessary for the most accurate description of solar resources for a certain application. In locations where there is very little variation between years, a shorter measuring period on the local scale could be sufficient and vice versa. An analyst might make use of this knowledge to boost their trust in a data set that is necessary for a certain analysis by which they will participate.

The chapter covered two primary methods for analysing variability: spatio-temporal variability and hyper-temporal variability. Spatio-temporal variability describes the spatial variation of solar insolation on an annual basis as well as the spatial variability of solar insolation every month for an entire year. Hyper-temporal variability examines the variation in solar insolation over a longer period than spatio-temporal variability does. Similarly, the hyper-temporal variability study investigates how changes in solar insolation occur over time and throughout the varying seasons. Because of the hyper-temporal variability research, we were able to assess variability on a micro-temporal scale, such as on an hourly, daily, weekly, and monthly basis. Similarly, we were also able to measure variability on a hyper-temporal scale.

It was previously believed that the slopes and orientations of the terrain in the Kalaburagi area, which are responsible for major fluctuations in the city's eastern half, were the primary regulators of the solar insolation shifts that occurred there. These studies demonstrated that the solar resource is subject to a broad spectrum of variation, with values ranging from very insignificant to very considerable at any one time. Utilizing directional spatio-temporal gradient variability was another objective of this work, which attempted to map the variety of natural and semi-natural landscapes in the Kalaburagi area. This method determines the relative heterogeneity of each pixel region by calculating the long-term spatial and temporal variability of each pixel.

The current portion of research allowed us to investigate the possibility of spatio-temporal and hyper-temporal approaches for analysing solar insolation variability. These approaches are well known in geospatial sciences for precise variability analysis, but they have been applied to solar insolation data from the Indian subcontinent for the first time. The findings also revealed a significant number of unexpected outcomes. Analogous investigations can be quite beneficial for datasets derived from in-situ ground measurements.

Chapter 7*: *Feasibility Analysis an Energy System

Solar power has quickly emerged as a significant contributor to the energy supplies of a growing number of nations. In this vein, India has made strides over the last several decades to maximize the immense potential of the country's renewable energy resources. Support for these intelligent solar grid networks gives us the potential to maximize photovoltaic energy utilization, which is an initiative that is currently being undertaken by the energy sector to develop decentralized energy solutions that are based on renewable energy resources. Because smart grids are intelligent networks that use the capabilities of information-communication technology and sensor networks to iron out all of the flaws in the traditional system, they are capable of handling all types of problems associated with the typical power supply or grid network. A smart grid integrates renewable energy or solar energy systems into the grid almost flawlessly by conveying how much input the renewable energy resources would bring to the grid and changing system variables like voltage and utility power. Consumers save money, and the number of dangerous air pollution from power generation is reduced as well. The first part of the work familiarized the reader with smart grid concepts as well as detailed explanations of smart solar grid networks, and the second part of the work assimilated all information about smart solar grids in a tabular list format to evaluate and compare the available technologies.

In the latter half of the chapter, the methodologies for doing feasibility analysis were thoroughly covered. The feasibility analysis was performed in three settings for the pilot study area, using the preliminary knowledge of feasibility. Technical, economic, and operational feasibility are the three settings of the feasibility study. Each parameter was given the appropriate weight. Payback analysis, return-on-investment (ROI), and net present value were the elements that controlled the economic feasibility. The cash flow charts were designed to examine each of these variables in great depth. The final feasibility matrix was created using all these characteristics, limitations, and considerations. The feasibility matrix could determine which technology should be chosen for photovoltaic installations. The weighted score for each of the technologies under consideration was provided by the feasibility analysis matrix. The weighted score was utilized to determine which technology should be deployed. In our research, we discovered that technology 2 outperforms the other two technologies. As a result, we chose technology 2 for the photovoltaic system installation.

The findings of an economic and technical feasibility analysis for the same project were simulated using a software tool called Hybrid Optimisation of Multiple Energy Resources (HOMER). The full explanations of everything were briefed, starting with the input and ending with the final output. The usual system architecture of a typical photovoltaic system was also depicted.

Finally, the study conducted a thorough SWOT analysis of the project. The final results were given in the form of a SWOT analysis after the suitable analysis. According to the author's knowledge, this matrix was one of the earliest novel approaches for performing any feasibility investigation.

The decision of whether to proceed is not always straightforward. In this sense, the current study contributed to a better understanding of the dangers and benefits of migrating. The feasibility study interpretations detailed the many possibilities, ramifications, strengths, and shortcomings. The feasibility study recommended that the project's scope be narrowed to determine the best possible scenario(s), with an investigating function to answer the question *"Is this a viable venture?"* These blueprints outlined the steps necessary to take the project from *"idea"* to *"reality"*. The research also assisted in determining the *"best"* option among the several situations. Feasibility studies do not turn positive or negative overnight. The research aids in determining the trade-off between the dangers and benefits of migrating.

Chapter* 8**: ***Synthesis and Transitions in Energy Systems and Policy

This chapter gave a comprehensive synthesis and conclusion of the research's primary findings, as well as its practical implications for the research community in the future.

8.3 Summary of Conclusions

Solar energy is one of the most environmentally friendly renewable energy sources since it is non-polluting and can be easily turned into electricity using photovoltaic (PV) systems. It is a valuable source of energy for the environment because it helps to lessen the dangers of climate change by reducing fossil fuel emissions and hazardous greenhouse gases, particularly carbon dioxide. The current study began with the goal of promoting green energy and inspiring people to use/install solar PV systems to stimulate the use of PV systems to help alleviate the massive electrical grid demands. With the ever-increasing need for alternative energy sources, every site should be pre-assessed for solar potential and suitability for PV systems. Shade from barriers, current local weather conditions, accessible rooftop surface area, and the surrounding environment are all major contributors to insolation losses.

To resolve these issues, the research presented in this thesis developed a precise methodology for modelling and assessing solar radiations, data processing techniques for processing modelled datasets, solar map generation method, solar insolation estimation technique for photovoltaic yield estimations, spatio-temporal and hyper-temporal variability analysis method for solar radiation, and feasibility analysis techniques. The major accomplishments of this thesis include the methodology developed to estimate model solar radiation, data processing techniques, solar map generation methods, hyper-temporal variability analysis method, feasibility analysis method, and gross solar PV electricity generation potential on rooftops for the pilot study area. Solar radiation models comprising one or more of the following characteristics, namely latitude, topography, and DOY, were developed in this study. For the pilot research region, there were undeniable gains above the traditional latitude-topography model. The work created an integrated methodology for modelling and mapping solar energy potentials in metropolitan areas every year, which was the inspiration for our current effort. In the case study area, a total of twenty-five buildings were chosen and assessed for their solar PV suitability. The software tools were tested for their ability to execute solar assessments on highly detailed 3D models and predict hourly solar PV power generation.

The capabilities of geospatial technology for solar energy modelling and feasibility studies for mapping resources on geographical and temporal scales have been validated in this study. Regional energy planning will undoubtedly benefit from the geographical database of resource availability and demand. It also provided the tools for detecting and measuring the elements that influence the amount of solar energy that is accessible. It also helped to increase the database's versatility, which is used to make many judgments. Additionally, it has been concluded that *"Greed"* has always been the greatest motor for the human species, although this is contingent on the fact that it is functional. It is now more important than ever to diversify our energy sources. The future will be put in jeopardy if we do not take immediate action to reduce or do away with our dependency on fossil fuels. Solar power systems are presently the sole technology that can be implemented in any of the world's most destitute developing nations. Because of this, we are of the opinion that within a few years, solar energy will be the first sustainable source of energy if researchers keep working diligently to improve the processes.

8.4 Recommendations to tackle the Energy Crisis in India

Following thereafter, actions must be done to improve energy supply on the one hand, while controlling demand on the other, by doing the following:

1. Efforts should be made to discover and implement new and renewable energy sources.
2. The government provides incentives to the private sector to invest in electricity generation projects.
3. To enhance India's energy status, several public sector undertakings and other organizations will be developed.
4. Efforts to increase fuel economy in the transportation and other industrial sectors to save energy.
5. Procedures for keeping Petroleum, Oil, and Lubricant (POL) usage to a tolerable level.

8.5 Prospects for Future Work

The potential directions that might be taken by the future study can be broken down into the following categories:

Increased precision in the measuring standard

As a direct consequence of this, measurement errors are sent to the modelling uncertainties and added to them. Models that are produced or confirmed with the most recent measurable data may only be as accurate as the measurement uncertainties that are currently present. To put it another way, the accuracy of any model, regardless of how exactly it may be, is limited by the data on which it is built. To develop models that are based on data that can be relied upon more, it is recommended that high-quality on-site radiation measurements be carried out in at least a few different sites.

Some modest suggestions for future research in a comparable setting may be found within the vast amount of research:

1. Optimal distribution network energy line routing to minimize energy dissipation.
2. Grid location and PV installation station suitability study.
3. Hourly power generating potential and a cumulative estimate of hourly solar PV electricity output, as well as a percentage offset for actual electricity usage.
4. Assisting policymakers in the installation of low-power appliances and making recommendations for future PV technology upgrades to boost energy generation.

There are opportunities to use a variety of tools to create a 3D model of the terrain surface to improve the interpretability and visibility of the terrain surface.

8.6 Transitions in Energy Systems and Policies

The switch to renewables affects buildings, transportation, industry, and electricity. Renewables may enhance cities' air quality, services, and living spaces. Cities use 65% of global energy and produce 70% of man-made carbon emissions, making them important to a low-carbon economy. Renewable Energy in Towns examines 3,649 cities' capacity to grow up renewables by 2030. Cities can boost renewable energy use cost-effectively. The study suggests three areas where cities may act: renewable energy in buildings (for heating, cooling, cooking, and appliances), sustainable transport (electric mobility, biofuels), and integrated urban energy systems.

This study showcases best practices from across the world to help communities shift to renewable energy. It explains how municipal leaders and administrators may design, regulate, finance, and operate urban infrastructure to hasten the transition to renewable energy. Energy storage has long been considered the next big thing, but it's been sluggish to deliver. A renewable energy-powered system can't operate without storage. But technical developments are modest, and there's no universal answer. Different technologies have their role as the application balances power storage duration, deterioration, discharge speed, and prices. Intermittent solar and wind energy is transforming power systems throughout the globe as these technologies dominate generation. This change in power mix offers cleaner energy but complicates the system. Higher renewable penetration reduces energy reliability. The electricity system must be able to manage greater stress when the sun shines and the wind blows. Active energy management and a fast grid are needed. Demand continues to rise, demanding more electricity, more efficiently, while meeting tougher environmental criteria.

Renewable energy isn't immune to the economic chaos. Construction projects struggle to find equipment. It's tougher to relocate operating and maintenance teams.

Low electricity costs will pressure profitability. The drop in oil prices will affect oil and gas corporations' capacity to invest in renewable energy, which they just embraced. Reconsidering urban infrastructure, information systems, as well as social and environmental justice, is necessary in order to develop a sustainable urban energy system. We provide prospects to *"green"* urban development and critical theoretical and practical research problems for sustainable cities. We advocate research, practice, and a new set of data-intensive indicators to attain sustainable objectives. Structures that generate renewable energy, store any extra production, and then feed it back into the regional grid when the time is right (either because of a demand for it or because of the price) would make up an energy-efficient network.

In addition to structures and transportation hubs, urban areas include complex and disjointed food and water distribution networks. It's possible that many city-wide services might be better off under regional or distant oversight. Opportunities to improve infrastructure, such as public transit and high-density housing, abound throughout Asia's megacities. It is likely that walkability, quality of life, and carbon emissions would all suffer if urbanization necessitates the mass transportation of people, goods, water, and energy. Though a higher population density might reduce the energy needed to power transportation, that need could not be met by renewable energy sources incorporated into the city alone. There are theoretical and practical issues with scale that are brought up by the interconnectedness of physical and administrative infrastructure. Economic, technical, behavioural, and legislative challenges plague efforts to reduce urban carbon emissions.

Nevertheless, there are a number of quick steps that may be done to start addressing these issues, including the following: *(i) specifically through the lens of the positive advantages that environmental justice brings, clean urban settings should be economically valued, and this valuation should be used to invest in underserved neighbourhoods, (ii) accounting methods for carbon and water should be standardized in order to boost resource efficiency now and pave the way for a transition to resource and pollution markets in the future, (iii) Combining congestion pricing with fee-bates, which are taxes connected with polluting cars that support the purchase of clean vehicles, is an effective strategy for improving the air quality in metropolitan areas and reclaiming city centres for walkers and social spaces.* In the meanwhile, we are still faced with the following important research problems.

Large Asian cities have seen rapid population growth and rising resource demands in recent decades, making metropolitan areas more polluted and resource-intensive than ever before. This is due to the increased population density often seen in metropolitan areas. Reversing the current trends in Asia requires a redistribution of resources and a renewed focus on the quality of life for all people. Sooner or later, similar patterns of development will be seen in other locations. To what extent city management can move to a paradigm in which environmental justice is not merely an occasional response to crises of inequality and in which we can reap the benefits of proactive and inclusive planning, design, and operation is arguably the most important question concerning the sustainability of urban areas. Access to renewable energy sources, enhanced energy efficiency technology, and carbon-light modes of transportation is critical. This is crucial for low-income groups since they are more likely to not own their homes and to have longer commutes. Cities may be made more liveable by addressing issues such as the safety of their streets, the cleanliness of their air, and the attractiveness of their parks.

These problems will need technological, social, behavioural, and commercial advances. Cities constructed around resource needs or personal cars failed to generate sustainable areas. Physical form, function, and ultimately justice and fairness provide a new way to low-carbon sustainable cities. Accelerating innovation and implementation can help cities decrease greenhouse gas emissions and become more liveable and fairer.

We don't inherit the earth from our parents; we borrow it from our children

Antoine de Saint-Exupéry

The manufacturer's authorised representative in the EU is Springer Nature Customer Service Centre GmbH, Europaplatz 3, 69115 Heidelberg, Germany. If you have any concerns regarding our products, please contact ProductSafety@springernature.com

Printed and bound by CPI Group (UK) Ltd, Croydon, CR0 4YY

30/09/2026

02225304-0003